Early praise for *Programming Sound with Pure Data*

This book covering Pure Data is pure fun. Where else can you learn how to make lightsaber sounds with code? It's a very nice intro to Pd and basic sound design and focuses on practical things you can use in your own apps.

➤ Jack Moffitt, senior research engineer, Mozilla

After reading *Programming Sound*, I had so many ideas running through my head that I couldn't sleep. I learned not only how to synthesize amazing sounds (think oceans, wind, wineglasses, and laser swords), but also why and how to dynamically incorporate those in both web and native mobile applications. Thank you, Tony!

➤ Zebulon Bowles, freelance musician and developer

Are you a sound designer? Working on a game audio project? Still looking to bring your projects to the next level? Look no further. Pure Data is your new secret weapon. And this is the book that will help you to get the most out of this powerful open source application. Tony delivers!

➤ Benjamin Lemon, composer, sound designer

Hillerson makes the depth of Pure Data accessible, and in the process teaches the basics of digital sound synthesis. Useful, fun, and highly recommended.

➤ Ben Price, senior developer, DreamQuest Games

Programming Sound raises the bar for what defines a good audio experience in software. This book will help you see application development through an entirely different lens. It arms you with the tools necessary to provide another avenue to engage and delight your audience.

➤ Dan Berry, iOS developer, Tack Mobile

Programming Sound with Pure Data

Make Your Apps Come Alive with Dynamic Audio

Tony Hillerson

The Pragmatic Bookshelf

Dallas, Texas • Raleigh, North Carolina

Many of the designations used by manufacturers and sellers to distinguish their products are claimed as trademarks. Where those designations appear in this book, and The Pragmatic Programmers, LLC was aware of a trademark claim, the designations have been printed in initial capital letters or in all capitals. The Pragmatic Starter Kit, The Pragmatic Programmer, Pragmatic Programming, Pragmatic Bookshelf, PragProg and the linking *g* device are trademarks of The Pragmatic Programmers, LLC.

Every precaution was taken in the preparation of this book. However, the publisher assumes no responsibility for errors or omissions, or for damages that may result from the use of information (including program listings) contained herein.

Our Pragmatic courses, workshops, and other products can help you and your team create better software and have more fun. For more information, as well as the latest Pragmatic titles, please visit us at *http://pragprog.com*.

The team that produced this book includes:

Jacquelyn Carter (editor)
Potomac Indexing, LLC (indexer)
Candace Cunningham (copyeditor)
David J Kelly (typesetter)
Janet Furlow (producer)
Juliet Benda (rights)
Ellie Callahan (support)

Printed in the United States of America.
ISBN-13: 978-1-93778-566-6
Printed on acid-free paper.
Book version: P1.0—January, 2014

Contents

Foreword

Increasingly, new arts and media projects require some programming skills. Studies in product design, interaction design, and user experience draw not only on the traditional fields of sound, graphics, and haptics, but also on practical skills with tiny embedded single-board computers like the Raspberry Pi and Arduino, on high-level rapid-prototyping languages, and on knowledge of browser technology and languages such as Python and C. One of the most popular and powerful rapid-development tools for audio-signal processing is Pure Data, Miller Puckette's wonderful visual programming tool. With its vibrant free-software community, and through the work of Peter Brinkman, Hans Christoph Steiner, and many others, it has emerged as the number-one choice for bridging the creative arts and engineering sciences.

As new university courses continue to expand, teaching sound design, sonic interaction, and sound arts as well as the traditional computer music offerings, the challenge of learning Pure Data in just one semester has arisen. Miller's book, *The Theory and Technique of Electronic Music*,[1] and my own, *Designing Sound [Far12]*, offer advanced treatments of substantially theoretical fields: electronic music and procedural audio, respectively. Each has practical elements directed at mastering these. Their use of Pure Data as a vehicle to convey more advanced academic material makes learning it something of a side effect. I like to recommend Johannes Kriedler's *Loadbang* to my freshman students since it is a great cookbook of "patch and go" recipes from which one can learn much by observation. Between these lies something of a gap, and beyond my own lecture notes I have often wanted to suggest a book that sets out to teach Pure Data as its primary objective. Tony Hillerson offers just such a resource here—a practical, step-wise, guided journey through mastering the basics of Pure Data and applying them to game applications using the C/++ libpd libraries.

---

1.   http://crca.ucsd.edu/~msp/techniques/latest/book-html/

You cannot learn programming from a book, but there is something valuable in this book that will motivate you and set you on the right course. Only at the end will you understand why the path is made by walking, and why I endorse this particular journey. A key factor is that the author has clearly trodden the road too, and speaks from personal experience, not a collection of data sheets and theoretical texts.

In *The Matrix*, Neo plugs in and downloads kung fu. If only learning could be like that. Programming is very much like kung fu. Indeed, it is no surprise that hackers talk this way about their skills. That is to say, it is all about experience. You learn to program by doing it. To some extent motor memory and ineffable symbolic knowledge play a part, and with time one learns to translate ideas and intentions into code without really thinking about the mechanics.

In the end this is a long journey. The ten thousand hours needed to become a master appear daunting in this age of impatience, and in the interdisciplinary digital arts where demands are made upon your skills in mathematics, planning, physics, graphics, psychology, networks, and electronics, it seems an impossible learning curve for which even two or three lifetimes are not enough.

But the human brain is amazing. Given a motive, an itch to scratch, the right seeds to grow, and a little sunlight of encouragement, we are able to extrapolate, create, and dream our way to extraordinary accomplishments, even with meager teaching resources. The voice of a good teacher is ever present in this book, giving solid formative steps and encouraging you toward exploration. Self-learning and the mathematical arts go hand-in-hand. The autodidactic approach to programming requires curious playfulness that can be brought out only through clear and quite simple objectives designed to be engaging; evoke personal, emotive goals; and subtly suggest further work. Activities like making a browser-based game are perfect for this project. Be prepared to have a lot of fun and develop a healthy addiction to code.

Andy Farnell
Computer scientist, author of *Designing Sound*
London, England, October 2013

Acknowledgments

After coauthoring a book in the past, I told a lot of people, loudly, "never again!" However, discovering Pure Data a few years ago gave me a chance to converge two of my happiest pursuits: working with sound and programming. When I discover something, I like to tell people what I've learned, and I was so excited about Pure Data that I decided to break my word and write a book again after all. I saw a gap in the literature about Pure Data, with not a lot of information for the beginner who wants to work with sound in mobile and web applications. I hope this book fills that gap.

Enthusiasm is enough to get started with when writing a book, but it's not enough to make the book a good one. I've needed the help of many other people: First of all, the people who've read it and given feedback and found errata during the beta. Then the reviewers, who helped immensely by offering feedback and advice, finding bugs, and pointing out algebraic errors. I've heard math in general well spoken of, but have yet to look into it. Thank you, all! The reviewers are listed here in purely alphabetic order (by first name):

Anthony Hoang	Ben Lemon	Ben Price
Björn Eriksson	Dan Berry	Ian Dees
Jack Moffitt	John Athayde	Mike Riley
Scott R. Looney	Zebulon Bowles	

I'd like to thank Andy Farnell, who graciously provided a foreword. From his excellent *Designing Sound [Far12]* I not only learned a great deal about designing sound with Pure Data, but also got solid grounding in the principles of sound that I had only intuitive or empirical knowledge of before. I heartily recommend that anyone wanting to learn more about how sound works and how to work with sound pick up his book after reading this one.

Next, I'd like to thank the publishers, Dave and Andy, my editor Jackie Carter, and everyone at The Pragmatic Bookshelf. I've always thought of the Prags' mark as a sign of excellence in our field, and to be able to write for them is an honor indeed. The tools they provide make writing a technical book so much easier than had been my experience in the past. In particular, Jackie's tireless efforts at banging my pedantic, pedestrian, prosaic, parenthetical, run-on sentences into some sort of presentable shape are particularly appreciated.

Thanks to everyone at Tack Mobile, and especially my partners John Myers and Juan Sanchez. I'm proud of what we've built so far, and I hope my interest in programming sound and sound design can be one aspect of the great software we continue to build as a company.

Finally, thank you to my wife Lori and our boys Titus and Lincoln for sparing me the time to write this book. Titus and Lincoln, I hope we can share some interest in programming or making music or both, but if not, I know whatever your interests and pursuits turn out to be, you will be great at them. Lori, I know I won't be able to interest you in programming, but I love you anyway. And to Diāna, here's a hopeful Laipni lūdzam mūsu ģimenē![1]

Thank you, all!

---

1. Thanks also to Helene for help with the translation.

Preface

I love sound. That may seem a little strange; a lot of people would say they love music, but I love sound—*including* music. I group them together, and I find that the line between them is blurry. There are interesting rhythmic and even melodic and harmonic elements in everyday sounds, if you listen for them. On the other hand, elements of music when taken out of context can lose their musicality and sound like natural events.

I hear little things all the time, which can be either good or bad. If something's rattling in the car it drives me *crazy* and I have to stop the car and make the noise stop. That makes *my wife* crazy, but I can't help myself. That's why it's so important to me to have sound done well in digital experiences, where I spend most of my time. When I say "digital experiences" I mean anything happening on a computer: web pages, web applications, native applications, mobile apps, kiosk apps, art installations, and, of course, games in any form.

Sound, and now I mean sound effects as distinct from music, is tricky to get right in digital experiences. Sound is very rarely a necessity. Even in games, where it's expected as a part of the experience, it's rare that the user *needs* sound to play. There are usually controls somewhere to turn sound off, but you'll never find a control to turn off the screen or mouse or touch interaction.

On the other hand, playing a game without sound takes away an important dimension of the experience. It's flat, boring, unconvincing. If there's not a little *ding* when you collect a coin, it's not satisfying. If you're running through the forest and you don't hear forest sounds, it's the opposite of immersive—it feels like merely looking at someone running through the forest on a little screen instead of *being there*.

Not only games can benefit from sound in this way, though. A few apps I've used struck the right tone, if you'll pardon the pun, with some tasteful and, above all, *meaningful* sounds as part of the experience.

This book is not able to teach you good taste, sadly. But, if you decide this book is for you, you will learn how to gain understanding and control over the sound you decide fits the digital experiences you create. You'll learn how to design and create the sound you want to hear.

Who Is This Book For?

I make a few assumptions about you, the reader. First of all, I assume you're as interested in the audible experience as I am. You may be a programmer or designer. You're probably interested in what you can learn about the technical details of getting sounds into a web or native app, but it's OK if you're just interested in learning about some technical approaches to designing sound.

I assume you've dabbled with sound design in the past. It was most likely by digging around in a sample library for the right premade sound and maybe editing that sound a bit to get what you wanted out of it. Maybe you're intrigued by the idea of doing some professional sound design in the game or movie industry, or maybe not, but for now you want to know how to gain more control over your ability to enhance digital experiences both native and on the Web.

You may also be a musician. As I mentioned, this book isn't about music in digital experiences; it's geared toward sound effects. However, all of the skills you'll learn here are easily applicable to musical experiences and musical apps, and at the very least they'll give you a better understanding of how to think about sound in music.

In short, I assume you have little to no experience with anything but dabbling with sound, but you're interested and want to learn more. I would be extremely happy if this book helps out those independent or small-shop developers and designers who have to wear many hats during the day and don't have the budget for a dedicated sound team.

What This Book Covers

This is a practical book for people interested in digital sound design for the Web and native apps. We'll focus on using one software application, Pure Data. We'll go through the steps of creating two sets of practical examples: sound-effect scenes that can be controlled dynamically or output to a file.

The first set will illustrate a few different sound-synthesis methods, and the second will show how to work with existing sounds, like you might get from a professional sound-effect library.

After getting used to using Pure Data and creating these examples, we'll look briefly at how to get the sounds we've made out of Pure Data and into a usable format, and then quickly discuss some sound-production topics.

We'll then take a brief interlude to discuss the user experience (UX) of sound on the Web and in native applications.

Finally, we'll consider two premade projects, a web project and a native application. We'll design sound effects for both of these, taking into account what the best UX for each is and what you've learned of sound design, and consider the possibilities of dynamic sound in the native applications.

We'll wrap up by considering what you've learned so far and a few possible directions you could take next in your journey to becoming a sound designer.

What This Book Doesn't Cover

Sound design is a deep and rich field. It grew up with the movie industry and is an essential part of the game industry. There are also many different approaches to designing sound within those areas. This book is an introduction to sound design for digital experiences on the Web and in native applications, which could include games.

We've chosen a pragmatic approach to introducing this subject; one that focuses on building practical examples using synthesis with a procedural sound application, which is only one of many possible ways to design sound. Here are some other important topics that aren't in scope for this book, but you may want to dig into next:

- This book does not cover capturing or recording sounds for analysis or sound production. Audio engineering, field recording, and Foley studio work are all good things to have under your belt as a sound designer, but we won't discuss them in any depth here.

- Analysis of recorded sound is also beyond the scope of this book. To reproduce a convincing effect it's very useful to know what's going on in a real-world example instead of guessing. We'll only have space to do a very rudimentary, intuitive analysis of some sounds as we build our examples, but as an area for later study, sound analysis would be very beneficial.

- There is a lot of math behind describing and reproducing sound. Algebra. Trigonometry for describing oscillators and geometric waves. Calculus for, among other things, Fourier analysis. That math is good to know, but beyond the scope of this book.

- A deep understanding of the science of sound is extremely helpful to the sound designer. In the field of physics, material physics, fluid dynamics, acoustics, and many others will help you understand from first principles how sound is made and how to reproduce it. Understanding the math used to express the physics in these areas will be helpful too. All of our practical examples will be intuitively designed, and we won't take the time to dig into the science of why the sound works the way it does. A great next step would be to start to understand the underlying causes of sound.

- Psychoacoustics, or the study of how humans process and perceive sound, is important to understand as a sound designer. There are some surprising things to learn from the field that can help you become a better and more efficient sound designer, but we won't have time to dig into those here.

Where Will You Be When You Finish This Book?

When you've read this book you'll have had the experience of creating a set of sound effects using Pure Data, ready for inclusion in a dynamic sound application or for creating a static sound file. You'll have a practical understanding of sound-synthesis techniques and a general understanding of how different fundamental components of sound can be used to produce sounds anywhere, from those we hear every day to those we'd hear only in a futuristic science-fiction world.

You'll be ready to take the next steps to understanding and growing as a sound designer, whether that be practicing what you learn here, building sounds for games and apps you'll be making, or digging in deeper to the math, physics, and other fundamentals of sound design.

Online Resources

On The Pragmatic Bookshelf's page for this book there are some important resources.[1] There are code downloads containing all the Pure Data patches you'll see in this book, as well as the source for the web, Android, and iOS projects. You'll also find feedback tools such as a community forum and an errata-submission form where you can recommend changes for future releases of the book.

---

1. http://pragprog.com/book/thsound/programming-sound-with-pure-data

CHAPTER 1

Introduction

This is a book about programming sound. Just as in any other programming book, in this book we'll cover the technical skills and tools that will enable you to tell a computer how to do something—namely, make sound. But just as with any other programming language, good design is important. I don't mean visual design; I mean careful thought and practice concerning "how things work." So I don't want you to think about yourself as a sound *programmer*; I want you to think of yourself as a sound *designer*.

Sound design is a practical art. Sound designers draw on an understanding of physical phenomena, technical knowledge, and intuition to create sound experiences. This book is about giving you the technical knowledge and providing some practical examples that will help you grow your understanding of how sound works, which will enable you to design and program the sound you want to hear in the digital experiences you create.

I have a theory: since humans rely primarily on their visual sense—way more than the other senses, in fact—experiences that involve the other senses have a greater chance of creating an unexpectedly good experience. Since the visual part of most digital experiences is so prevalent, it's surprisingly more realistic when the other senses are involved.

Digital experiences aren't heavily tactile yet, besides the mundane input methods such as typing or mousing, or maybe playing on a game controller shaped like a guitar. There are interesting new input methods like the Nintendo Wii Remote, Microsoft Kinect, Leap Motion device, and so on, but there's nothing mainstream that provides tactile feedback. And since Smell-O-Vision has yet to take off,[1] the audible experience is the best field within which we

---

1. http://en.wikipedia.org/wiki/Smell-O-Vision

can create that extra, sensory magic in addition to the visual to immerse the user in the experience.

Before we talk about Pure Data, the language and tools we'll be working with throughout the book, let's discuss a little about sound design and what it means. Then we'll take a look at Pure Data from a high level to get oriented before we dive in and start using it to make sound.

Getting Started with Sound Design

Let's talk a little about sound design. This is a general overview to give you an idea of the scope of the field, and to provide a context for the more technical chapters to follow.

The Sound Designer's Goals

The digital sound designer may have any of a number of goals when creating a sound. Sometimes these goals overlap, but in general this is what we want to achieve in an experience using sound.

Adding Audible Feedback

Adding audible feedback is a common use of sound in digital experiences. Such feedback could be a clicking sound when the user presses a button, a beep or bell tone when a task is complete, or a notification when a message arrives.

Fulfilling Expectation

When some event takes place onscreen, it can carry with it the expectation that if this event happened in the *real world*, a sound would be part of the event. In a game, a weapon is fired, a rock falls to the ground, or something is stretched or struck or thrown. The user expects things to sound a certain way depending on a lot of factors, such as the realism of the experience and conventions for similar experiences. Gauging these expectations and designing to meet them are goals for a sound designer.

Communicating a Mood

Background music in a game is a clear example of sound for communicating a mood, but it needn't be only in a game. Think of the startup sound of your computer system of choice. The Windows and Mac startup sounds are designed to signal an event, but also set the mood for stepping into a fresh, clean session.

Creating Immersion

Immersion may be related to communicating a mood or fulfilling expectations, such as when getting ambient environment sounds right in a game, but the goal of immersing the user goes beyond those to creating a believable, emotionally involving experience. The goal is to make the user forget about the outside world.

Prompting an Emotional Response

Creating an emotional response is related to immersion, but I call it out separately because there may be a goal to quickly get a response from the user, maybe with a musical stab, or the scream of a monster from an unexpected direction, or a sound cue signaling the successful completion of a task.

Types of Sound

Now that you have some possible goals in mind, let's discuss the kinds of sounds you may use to reach these goals.

Music

Music is an extremely powerful type of sound. The human response to music is a mystical one, and even if your interest lies more in creating other types of sound effects, gaining an understanding of music will help. This book will not cover music directly, in terms of musical theory or creating musical instruments, but a lot of the skills we discuss can be applied to musical applications.

Ambience

Ambient sounds are those that occur in the background. These sounds don't have to be continuous, but they often are. The effects are there to help with creating an immersive environment or to fulfill the expectation of a certain situation. If in a game there's a scene where the wind is blowing, the user will expect to hear wind, for instance.

Effects

Non-environmental effects, sometimes called *hard effects*, are those that occur in conjunction with an event in the foreground, or somewhere the user's attention is expected to be. Such effects include a button click, a door slam, and a laser blast.

The Sound Designer's Methods

Given those categories for effects, let's look at how sound can be created.

Sampling

The easiest way to get a realistic sound is to record it. This is called *sampling*. Using a digital recorder out in the field to record the sound of a stream or a jet flying overhead is a reliable way to capture a sound. Sampling can be done in a sound studio, too, which offers more control over the environment the sound is recorded in, but places limits on what kind of sound can be captured. You can see the tradeoffs between the two.

Sample Library

Recording your own samples can require a fairly extensive investment in space, time, and equipment. If you've done any work with sound in apps in the past, you've probably cut out a lot of work and bought a prerecorded sample library. There's a number of these professional sound libraries out there for different budgets, and a number of online services offer various samples a la carte.

Synthesis

Sound synthesis is the process of constructing sound from fundamental sound components—*sound waves*. It really is surprising how realistic sounds can be when built up from generated sound waves. The complex part isn't finding the right sound or environment in the real world, or creating a plausible situation in a studio, but rather understanding how sound is made well enough to be able to re-create the sound from scratch using a synthesis environment. This technique is this book's focus.

Building Sounds

The techniques described in the preceding sections are all time-tested and approved. I'm not making any claims that one is the right one—a mixed approach could be the best, and the needs and budget of each project will make one or more of them make sense and exclude others. This book is primarily about the exciting possibilities digital synthesis opens to sound designers, with the ability to both build sounds from scratch and make use of a sound engine inside the digital experience.

The biggest gain is that of extreme flexibility. If you need the sound of many different types of explosions, you could try to record many different things blowing up. If you instead create a sufficiently complex model of an explosion in a sound-generation application, you can create an infinite number of variations. It's even better if that model can be embedded in your application and react to different parameters controlling the explosion sound—much more useful than a directory full of explosion samples.

This approach mixes well with the other techniques I've described because the samples can be manipulated and controlled in the synthesis environment. That gives new life to your sample library and gives you the option to start with a sample that may be hard to synthesize. A few synthesis applications are already out there, with a varying amount of complexity, power, and price. This book focuses on a popular application called Pure Data, which is open source, free, stable, and very powerful.

In preparation for jumping in and making sound, let's take a high-level look at what kind of software Pure Data is and how you interact with it.

Introducing Pure Data

Pure Data, or Pd, as its users call it, is an open source, visual programming environment for building audio and visual experiences. It was created in the 1990s by Miller Puckette, and grew out of the ideas behind a previous creation of Miller's called Max, which is still available as a commercial product. The Pure Data website is http://puredata.info.

Pd is part of a class of programming languages and environments that deal with *procedural audio*, or creating sound from routines that generate or modify streams of numbers that will eventually be sent to hardware to produce audio that we can hear.

Pd Is Visual

Pd is a visual programming environment, which means while using it you don't write code as such, but instead manipulate visual objects on the screen, connecting them into a system that produces some desired effect. The visual objects are technically called *atoms*, but we'll use more specific names for them as we talk about them. Each atom has a particular job, whether to send a message to other atoms, hold a number, or configure and control an object from a core library of objects or third-party extensions and abstractions.

A Pd file is called a *patch* and has the extension .pd. A patch is a textual data file containing information about how atoms are connected, how they're configured, and sometimes data that they can read from.

When we open a patch in Pd a graphical representation of the patch is displayed in its own window. Atoms can be moved around the screen, connections can be made, values can be adjusted, and so forth. The work area inside the window is called a *canvas*. All Pd programming happens in this visual environment. We'll spend a lot of time going through patches, often looking at images directly taken from Pd patches. Early in the book we'll walk through building patches from scratch, and later we'll switch to a more descriptive, explanatory style, but remember that all the patches described in the book are in the code download in the pd directory.

> ### Joe asks:
> ## Can I Edit Patches or Put Them in Source Control?
>
> Although patches are simply text files, It would be a bit of a stretch to call them human-editable. They should be viewed as data files.
>
> Merging changes with those of a colleague is most likely impossible, so to work with Pd in a team environment, either have Git treat Pd files as binary files or deal with conflicts in another way. Making good use of abstractions will help here. Refer to Git's documentation or that of your favorite source-control program for more information.

One of the great things about Pd's visual design is that the graph shown in the patch (as in the figure here) is everything you need to know about how to reproduce the patch and understand the flow of the audio through the system. The information density of a picture of a patch is very high, and just looking at it makes explaining the patch's inner workings much easier than if we had only code to look at.

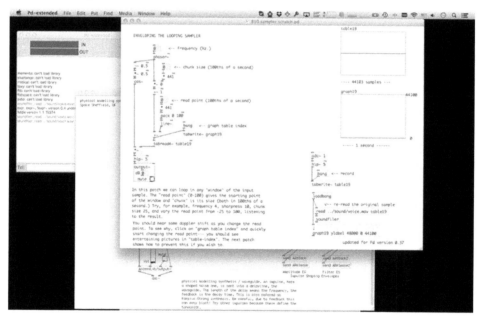

Pd Is Modular

Pd is great for learning how to build sounds because the visual programming environment makes it easy to communicate what's going on. It continues to be a powerful tool long after you start being productive, too, thanks to its modular design. It's easy to create abstractions at whatever level of complexity you need. These abstractions can themselves contain abstractions and so on, allowing the Pd programmer to gain a lot of control over and reusability of component pieces of a patch. The abstractions can be stored in a patch or in a separate file.

If Pd doesn't do something you need it to, you can write extensions in C. There are sound-processing extensions, visual libraries for art installations, interfaces between popular hardware controllers, and awesome stuff like Arduino and Raspberry Pi integration.

Pd Is Embeddable

Since in this book we're focused mostly on audio for web and native applications, one of the coolest things we'll do with Pd is embed Pd patches in native apps using the excellent libpd.[2] This will allow us to put dynamic, procedural audio directly into our apps and build a domain-specific language around the audio experience. When I first heard that was possible I nearly jumped up and danced. Not a pretty sight. That means Pd is a ready-made embeddable synthesizer engine for your native apps!

So, although Pure Data is only one of many ways to create and produce sounds that you design, it's a very powerful and useful tool to have in your sound-design tool belt.

Installing Pd

You can find installation instructions at the Pure Data website,[3] but which version should you install? There are two: the original version maintained by Miller, lovingly called Pd Vanilla, and Pd Extended. Pd Extended is easier to install, and comes with a number of third-party extensions, including sound-processing tools and effects. Patches created with extensions from Pd Extended will not run in Pd Vanilla unless the extensions are added to Pd's Vanilla's path, but otherwise they are almost completely compatible. As of this writing Pd Vanilla was at version 0.44, with 0.45 right around the corner. Pd Extended was using Pd 0.43 under the hood.

All of the patches in this book were created with Pd Extended, but I've kept away from any extensions that don't ship with Pd Vanilla—with one notable exception, the output~ object. I'll make note of it again when we first use it.

---

2. http://libpd.cc/
3. http://puredata.info

If you choose to use Pd Vanilla, there should be no issue with the patches in this book. If you use any of the third-party extensions that ship with Pd Extended, keep in mind that libpd, which we use to embed Pd in native apps, is a wrapper around Pd Vanilla. You'll need to track down and include any extensions you use alongside your patch.

Other Software

It's not necessary to have any other software for this book, but it might be useful. In a few cases I've included screenshots showing some audio-file analysis in my favorite audio editor, Adobe Audition.[4] A free, open source alternative is Audacity.[5] You're definitely going to want a full-featured audio editor, and either of these is a good choice.

In the final chapters of the book we'll go through two projects: a web game and a task-management app built for Android and iOS. Although you could learn a lot from the discussion in these chapters covering the Pure Data part alone, to get the most from the code you'll need a good text editor for the web code, Eclipse for the Android project,[6] or Xcode for the iOS project.[7]

Let's Get Started

You're beginning (or perhaps continuing) an exciting journey toward being able to design the sound you want to hear, which will give realism, emotional depth, and that extra dimension to make your apps and games come alive.

 Joe asks:
What's That Again?

Sound is an immense field. Unbelievably immense. This book is meant to be a practical, hands-on primer. I try to explain enough theoretical background to help make sense of what's going on when we come across a new concept, especially in the next chapter. If the theoretical parts leave you confused or asking more questions, use them as jumping-off points for further study—Wikipedia is a great place to start.

Now that you have a general idea of what we want to accomplish in this book, what the sound designer's goals are, and how we can use Pure Data to accomplish them, let's get started with Pd and make some noise.

---

4.   http://www.adobe.com/products/audition.html
5.   http://audacity.sourceforge.net/
6.   http://eclipse.org
7.   http://developer.apple.com

CHAPTER 2

Making Some Noise

Now let's get started programming Pd. In this chapter we'll cover a few important tools Pd has for generating sound, controlling it, and outputting the sound to the computer speakers. Most importantly, you'll learn how to make sound!

When you are done with this chapter, you will

- Understand some basic concepts about sound
- Connect things together to make a signal flow to the sound card
- Control the volume of sound we make
- Make sound!

Time to get hands-on with Pd.

Finding Your Way Around

We'll start with a new Pd file and discover a few things about editing. Open Pd, and you should see an image like Figure 1, *Editing a PD File*, on page 10.

In the top left are some meters to show input or output volume, in the top right is a check box marked *DSP*, and the rest of the window is taken up by a console with log messages. Notice that the DSP check box in the top right is unchecked. Leave it unchecked for now.

DSP stands for *digital signal processing*. This control toggles whether Pd processes and outputs sound to the operating system and on to your computer speakers. Always be thinking about what kind of sounds you're about to make and how to quickly make them stop. You can't easily replace your ears. If you are wearing headphones, check the system volume before you get ready to make sound with any application, including Pd. Also, remember that in Pd if whatever you do makes a sound you don't want to hear anymore, you can quickly jump to this main Pd window and uncheck DSP.

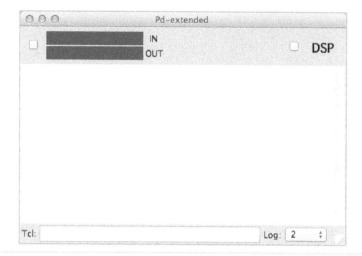

Figure 1—Editing a PD File

Creating a New Patch

Open a new Pd patch by using the File > New menu, or simply pressing ^N. A new window should open with a nice, clean blank *canvas*. Now open the Put menu and click Object. A dotted blue box is placed onto the canvas with a text field inside and ready for typing. If you move your mouse around the canvas, the object will follow it until you click or you press a key on the keyboard. Click on the canvas to place the object box in the top-left area of the canvas.

If You're on a Mac...

On the Mac you use Command (⌘) instead of Control (^) for any of the key commands you'll find in this book, but I'll give all key commands using the ^ key.

Now, if you click on the canvas away from the object, you'll notice that the dotted line around the box becomes red. This shows that this box doesn't mean anything to Pd right now. If you click back into the box, you're still able to edit it. You can also delete the box from your keyboard, but if you simply click the box, you'll find that the object is in edit mode and your delete key will delete characters inside the box instead of the box itself. The best approach is to band-select—click and drag your mouse around the object(s) you want to delete—and then delete it. Try that now: band-select the object and delete it. You should find yourself with an empty canvas again.

If you're going to use Pd for any length of time, you should get comfortable with a few simple key commands to put things on the canvas. The key sequence for putting an object on the canvas is ^1. Try this now: press ^1 and notice how an object is created wherever your mouse pointer is in the canvas. Once the object is on the canvas, you can drag it to position it where you want it.

Working with Objects

We now have an object on the screen. An object is in some ways much like an object in your favorite programming language: a building block that has some functionality. But, like in most programming languages, there's not a lot we can do with just an object; we need to tell Pd what kind of object we want. We do this by typing the object's *class name* in the box. This may be a little surprising if you've worked in a visual programming environment before, where you usually pick exactly the type of thing that you want from a set of menus, and then end up with complex property sheets to edit all sorts of parameters. Pd doesn't muck about with that sort of thing. It's visual, but it's simple. The following image shows the anatomy of an object.

Think of the class name of the object as something between a class, a function call, and a keyword in your favorite programming language. Another apt analogy is a shell command, such as cd, grep, or sed. The object box takes the name of the type of object you want to use and a list of optional arguments, just like a shell command, and just like a shell command, an object can vary widely in power and complexity. An object's outlets and inlets are the small bars at the top and bottom. Clicking and dragging from an inlet to an outlet will create a connection between the two. We'll look at connections and signal flow in just a second.

When typing in an object name, if you make a mistake about what you want, Pd will quietly tell you. Let's try it. Click inside the object you created with ^1 and type the word "nothing" into the box. Now click outside of the box. First of all you should see that the object's box has a red dashed line around it, just like the empty box did. Now switch back to the main Pd window and look in the console. At the bottom of the log messages you should see this:

```
nothing
... couldn't create
```

Pd couldn't create an object called nothing because it doesn't have an object type named nothing. If you ever try something that doesn't work but you're not immediately sure what's wrong, check the log in the console to see if Pd has more information for you. Also, remember Pd's help, which contains a lot of information about the objects that do exist.

Now we've covered how to create an object with both the menu and the quicker key sequence, position the object on the canvas, and tell Pd what type of object we want. We haven't yet created an object that Pd understands, though, so let's make Pd actually *do something* by creating an object Pd does know about: an oscillator.

Hello Concert A

An *oscillator* is something that moves back and forth between two states, and in Pd it's an object that we can use to make sound. We'll get into more detail in a second, but for now let's create one and wire it up to the computer speakers.

Making an Oscillator

Create an object, place it near the top of the canvas, and type in osc~ 440. Now create another object underneath the first and type in the name dac~. Move your mouse over the outlet: the small gray box on the bottom left of the osc~. You should see your mouse pointer turn from the rather quaint pointing hand icon to a ring. Drag from that point to the left gray box on the top left of the dac~ object, its left inlet. Now drag another line from the same bottom-left connection point on the osc~ to the top-*right* box on the dac~, its right inlet. You should have a canvas that looks roughly like this figure.

Let's hear what it sounds like. Check your volume to make sure it's low—say, in the lower quarter of full volume—and go to the main Pd window. Click the DSP button on, and you should hear a high-pitched tone coming from your speakers. When you're done listening, toggle DSP off.

Signal Flow

We've just created two objects and made Pd process and produce some sound. First of all, notice that both object names have a tilde (~) at the end of them. By convention, object names that deal with *scalar values*, or numerical values that don't make sound directly, have no tilde, and objects that have a tilde at the end deal with *signals*. That means that the object either produces or modifies a signal, a numerical stream of data that will get turned into a sound.

To understand what this patch does, let's start with the second object on the canvas, dac~. A dac~ sends a signal to the sound card. We'll dig in to how digital sound works a little later, but for now, know that DAC stands for *digital-to-analog converter*, and in this case it means the process by which your computer sound card makes sound. Whenever you hook a signal up to an object named dac~, you are sending a signal to your computer's sound card. The signal flows from an object's outlet into another object's inlet, and on until it reaches a dac~ and we're able to hear the signal as sound.

Now, back to the first object we created, the osc~. The object's name is 'osc~', which stands for *oscillator*. For now think of an osc~ as a sound-wave generator. If you're curious, the wave that it makes is a sine wave (a cosine wave, to be exact), but don't get too far ahead—we'll get into waves later.

As we saw, the little shaded boxes that are on the corners of the objects are the *outlets* and *inlets*. Inlets are at the top, and allow incoming data to control the object. Outlets are on the bottom, and send data out of the object. The lines that we drew from the outlet of the osc~ to the two inlets of the dac~ are called *connections*. Connections are fairly self-evident once you know you can draw them. You may notice a subtle difference between different connections: connections that carry signals instead of just numbers are a little thicker.

Getting Help

Pd comes with a nice help library that we can access and search from the Help menu. Even more useful, right-clicking an object will bring up a context menu that has an option for getting help about a particular object type. Try it now: right-click the osc~ object and read the help entry on the oscillator. Notice how it's broken into three sections—inlets, outlets, and arguments—and has links to related objects at the bottom. Whenever you want to know more about anything we cover in this book, the context help is a great place to start.

The Argument to the osc~ Object: Frequency

We passed an argument or parameter to the osc~: 440. Think of this like the object's constructor: we're creating an osc~ with an initialization value of 440. The value is optional, but by doing this we've created an oscillator with a *frequency* of 440 hertz. Abbreviated Hz, the unit hertz measures the number of times something that cycles between two positions does so in one second.

To illustrate frequency, think of a swing on a playground with a kid going full speed back and forth. The swing is an oscillator. If the kid's legs start out pointing forward, swing back as far as they can go, and then return to the point where they started all in the space of one second, then the frequency of the swing is 1 Hz. We also say that the swing has a *period* of one second. Period is the inverse of frequency.

An Audible "Hello World"

Now back to our osc~. It's oscillating 440 times per second. What does that mean? What's doing the oscillating? The wave that osc~ generates is oscillating at 440 Hz; to be more accurate, the signal the osc~ generates describes a wave with that frequency. The figure here can help you visualize this process.

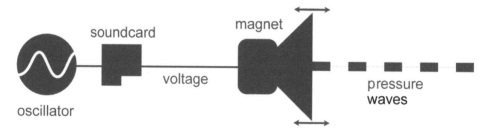

When the signal gets sent to the sound card and converted to something that can move your computer speaker or headphones, at that point the speaker is vibrating at 440 Hz. That speaker is attached to a magnet that pushes and pulls it back and forth, pushing air back and forth at 440 Hz, creating waves of high and low pressure in the air. Finally, parts of your ear are moving back and forth at 440 Hz. This is a simple description of how electronic sound works.

Now, why did I pick 440 Hz? It is a standard note in (most) Western music, called "Concert A," and often it's the note used to tune up an orchestra. In other words, you've just said an audible "Hello World" using Pd. Well done!

Things to Think About

Why are there two inlets to the dac~? See if you can guess, then turn on DSP and delete one connection by clicking and pressing delete. Listen to the difference. What do you notice?

What happens when you change the frequency of the osc~? How low can you go before you can't hear it anymore? How high? Hint: You may have to get up into the tens of thousands.

Controlling Volume

Now that we're making sound, remember that if things don't sound quite right you can quickly turn the sound off by unchecking DSP in the main window. It'd be nice to have a better way to control the volume right there in the patch, though. Let's look into a few ways to do that.

Using Operators

Since a signal is, at least when represented digitally, a stream of numbers, we can also multiply the signal by 0 to make it send a stream of zeros, or silence. To do that, we can use a *signal operator* in Pd. Pd has both *arithmetic operators*, which work on scalar values, and signal operators. You'll recognize each right away because they look like operators from your favorite programming language, like +, -, *, /, and so on. As always, if an operator works with a signal, it ends with a ~.

On the patch we already have open, with an osc~ connected to a dac~, select one of the connections between the two. As you hover your mouse over the connection you should see it turn into an X. This is how you know you're about to select a connection. It's helpful in case your patch gets a little cramped and you're not sure what you're about to select. Click to select the connection, then press Delete. Delete the other connection, too, while you're at it. Now you should have a patch that looks like the one in this figure.

Now press ^1 to create another object. Place the object in between the 'osc~' and the dac~ and call it *~ to create a signal-multiplication operator. Multiplying a signal by some value affects its *amplitude*, or the distance between the highs and lows of the wave—basically the volume of the sound.

To illustrate amplitude, let's go back to our previous example of a child on a swing, but this time consider two children next to each other swinging. One got a bit of a head start out to the school yard and is already going as high as she can. The other has just gotten started, and is only traveling a few feet forward and back. Suppose that they're both moving at the same frequency, meaning they both take the same amount of time to get from the maximum distance forward to the maximum distance backward. So the faster of the two is moving a greater distance but the full trip takes, say, one second. The other child is working hard, but at the moment only moving a few feet, although in the same amount of time, the same period: one second. The one traveling the greater distance has a lot more force in her swing.

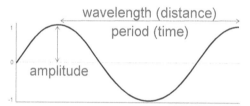

This diagram of a sine wave illustrates some of the concepts we've been talking about. The wave's amplitude is graphed on a scale between 1 and –1, with the wave centered at 0. The amplitude of a wave isn't necessarily limited, but when dealing with sound, a range of 1 to –1 makes sense when speaking about output volume, since speakers have a maximum distance they can be pushed out or in by the magnet controlling them. The amplitude of the wave is measured as the amount the peak goes above zero.

Measuring the distance between the wave's peaks gives us the *wavelength*, while measuring the amount of time between peaks give us the period. Again, the wave's frequency is how often the peaks of the wave occur over time. The preceding graphed wave shows how period and frequency are inversely related—if more peaks occur in a given time period, the period is shorter and the frequency is greater.

For sound, frequency affects pitch, and amplitude affects force and energy, which directly affects the volume of the sound we perceive. A 440 Hz sine-wave signal with an amplitude of 1 is louder than a 440 Hz sine-wave signal with an amplitude of 0.5. Of course, an amplitude of 0 won't make a sound. Let's try it.

Back to our working patch; connect the osc~ to the *~ and the *~ to the dac~, as shown here.

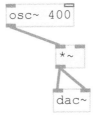

Now turn on DSP, and what do you hear? Nothing. As you can probably guess, the *~ operator needs to multiply two things together, and right now it's multiplying the signal by its default value of 0.

Using Messages

We'll use a new tool called a *message* to feed a value to the *~ operator. Instead of pressing ^1, this time press ^2. You could also use the menu Put > Message.

You'll notice that now you have a flag-shaped box instead of a regular square one. Place the message next to the *~ on the right. Now use ^2 to create another, and place that below the first. Double-click in the top box and enter a 1, and then put a 0 in the bottom box. Connect the bottom outlet of both messages to the top-right input of the *~, and you should have a patch that looks like this.

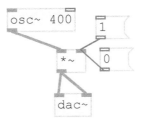

Before we can take the next step, we have to talk about Pd's two modes: *Edit mode* and *Run mode*. So far we've been working in Edit mode. In Edit mode your mouse can select, move, double-click to edit, and do all the familiar edit actions. If DSP is on and a signal is being sent through a dac~, then Pd will produce sound. If any of the controls you interact with have a user interaction associated with the button-like objects, sliders, etc., those interactions won't work, and clicking them with the mouse will only select them for editing or movement.

In Run mode, any of the Pd atoms that have user interactions will be active. Pressing buttonlike objects will activate them, and sliding sliders will slide them. If any of the atoms have no user interaction, as is the case with objects or connections, then clicking will do nothing. Of course, in Run mode Pd will output sound as long as DSP is on.

To switch between modes, you can toggle with the key command ^E. If you're in Run mode you'll switch to Edit mode, and vice versa. You could also access a toggle-menu item from Edit > Edit Mode.

It'd be nice if Pd made the current mode for a patch a little more obvious, perhaps something like Vim's INSERT status message at the very least. Instead, you can look for clues with the mouse pointer, which will be the hand pointer when you move it in Edit mode, and a regular system pointer in Run mode. Watch out, though—if you toggle the mode the mouse pointer will not update *until* you move it.

Back to our patch, toggle into Run mode by pressing ^E. Make sure DSP is on and your volume is at a reasonable level, and then press the message with the 1 inside it. You should be able to hear the 440 Hz tone. Now press the message with the 0 in it. You should hear nothing. Toggle the messages back and forth as much as you like and think about what must be happening. Imagine what would happen if you changed the value in one of the messages to 0.5. Try it and test your theory.

The *~ operator is multiplying the signal by the *scalar value* sent from either of the messages when you click them. Scalar values are just numbers, as opposed to signals. When the *~ operator is holding a value of 1, it multiplies the signal by 1, and the output flowing through the outlet on the bottom left is simply the signal. When the *~ operator is holding the value 0, the output is a continuous 0.

Messages are important controls in Pd. They are the first atom we've seen that has user interaction, and they can help you control what other atoms, like objects, are doing.

Using Controls

Let's briefly talk about a few other ways of controlling volume. First, a *~ object can take an initial argument, and that is sometimes a useful approach to controlling a signal with a specific value. Have a look at this figure and reason about what it does. Try it out if you like.

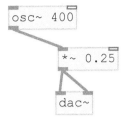

With the initial argument of 0.25 the signal coming from the oscillator is sent to the dac~ at one-quarter the volume.

Another approach is to use a *toggle*, which is like a check box. It works the same as our dual-message-box approach: when the toggle is checked it sends a 1 value through its outlet, and when it's unchecked it sends a 0. You can create a toggle with ^⇧T or from the menu with Put > Toggle. Create a patch that looks like the one shown here, and try clicking the toggle on and off.

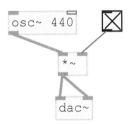

The last widget I'll introduce for volume is a slider, which will help us make a more flexible volume control. Sliders work as you might expect and let you output a scalar value between a min and max value. You can place a slider on the canvas with ^⇧H for a horizontal slider, or ^⇧V for a vertical slider.

Create a patch to look like the one in Figure 2, *Introducing the Slider*, on page 20, but *don't touch that slider yet!* This is important to remember: sliders have an initial min and max of 0 to 127. That will be *loud* if you slide the slider very far from the bottom. Right-click the slider and click Properties from the pop-up menu. Under the *output-range* header change the Bottom and Top values to 0 and 1, respectively.

Now try moving the slider control up and down to adjust the volume and notice how much more control you have than with hardcoded arguments to *~ or messages.

These controls in conjunction with a *~ operator are useful tools. You can use them in different places in a patch, especially as it grows more complex. Controlling the volume of a signal at different points in a patch is important,

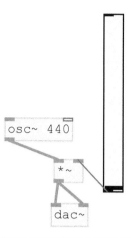

Figure 2—Introducing the Slider

whether the signal goes to a dac~ or passes through to another part of the patch. At the final stage of the patch when you want to send the sound to the speakers, it's nice to have a more powerful and ready-made object for controlling sound output. That's exactly what the output~ object is for.

The output~ Object

The output~ object is a set of controls included with Pd Extended for just about everything you want to do to control audio output. As you can see in the following patch, it has two inlets like a dac~, a volume slider, and a DSP toggle all in one. The DSP toggle here controls the global DSP toggle, which makes it a lot easier to shut down sound whenever you want. You can right-click and open the object to see more about how it works inside, but in essence it's a set of controls wrapping a dac~.

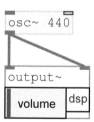

Most of the patches you see from now on will use output~ instead of dac~. Remember that since output~ is not included with Pd Vanilla, when we create patches for libpd, we'll switch back to using dac~.

Now that we can control the output volume, we're ready to make a more interesting patch using more than one frequency.

Working with Different Frequencies

Now let's have a little fun, and while we're at it, build a little more complex a patch. Make a new patch that looks like the one shown here.

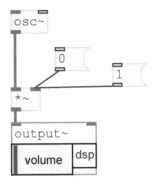

You can safely keep DSP on in this case because the osc~ doesn't have an initial frequency and it doesn't have anything connected to its inlet telling it a frequency to use. It will make no sound.

Using Simple MIDI Messages

Now, above the osc~ create an object and call it mtof. This object, mtof, means *MIDI to float*. Above that object, add a message with the scalar value 60.

 Joe asks:
What Is MIDI?

MIDI stands for *musical instrument digital interface*, and it's a spec that allows instruments—keyboards for the most part, but also other controller devices—to talk to each other.

You may already be familiar with MIDI, and to answer the question you may be asking, yes, you can easily hook into Pd with your MIDI device, to either perform or control. It's outside of the scope of this book, but if you're interested in it, keep it in mind as we start to build more complex patches. You may see ways that you can save yourself time by using an external controller to work alongside some of the widgets we'll use to control patches.

Your patch should look like the figure shown in Figure 3, *Patch with mtof Object*, on page 22.

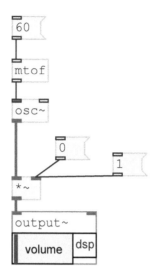

Figure 3—Patch with mtof Object

In the patch we're creating here we're not going to use an external MIDI device. We're just going to use the facility that mtof provides, which is to turn a MIDI note message into a floating-point number that an osc~ accepts as a value to its inlet for frequency. That way we can catch a glimpse of Pd's MIDI capabilities and more easily program the patch. The result is a surprise for now, but here's a hint: the number 60 corresponds to the MIDI note message for C on a keyboard.

If you want to test what we've got so far, switch to Run mode with ^E and press the 1 message connected to the *~, and then the 60 message. You should hear a tone, a little lower than the A we produced in previous patches. To stop hearing the tone press the 0 message. Now switch back to Edit mode.

We'll add two more objects. Add one right above the 0 message and name it delay 500, and connect it to the 0 message. A *delay* object, when activated, will wait a certain amount of time and then cause something else to happen.

Next, above the delay, create a new object and type in trigger float bang bang. You read that right. Once you stop laughing, just do it, and I'll explain.

Using Triggers

A trigger is an object that can cause multiple things to happen at one time. Now select and delete the connection between the 60 message and mtof, and instead connect the 60 message to the trigger. Connect the leftmost outlet of the trigger to mtof, the middle output to the delay object, and the rightmost output to the 1 message. Notice that there are as many outlets as there are parameters to the trigger. Objects in Pd can have a dynamic set of outlets, so if we had trigger float bang bang float, for instance, there would be four outlets. Before we move on, check that your patch looks like this figure.

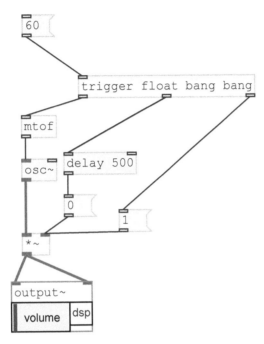

Now that we have something to look at, let's talk about this, starting with trigger. A trigger takes any number of initialization arguments and, when activated, sends its received input through its outlets. Whatever type of input activates the trigger will be sent along all of the trigger's dynamic outlets and converted according to the type of the outlet.

In our case the triggering event is the message sending its value, which is 60. The trigger receives that number and sends it as a float to outlet one. It converts the number to a bang and sends a bang to outlets two and three. A bang is a type of event that simply activates an object. It's like telling an object, "Do whatever it is you do now." It's like a little nudge to an object. So now when you press the 60 message, the trigger will send 60 as a float to mtof. It will now also send a bang to the delay and the 1 message.

Let's talk about the order in which the trigger sends these events, though, because it's subtle and important. The trigger sends the events right to left. That means that in this case the 1 message gets a chance to do its thing first, which is to send a message to the *~ object. When a message receives a bang it's the same as if you had clicked it. The delay receives its input second, and then the mtof.

Now that you see how a trigger works, consider the delay. If the delay is initialized with a value, which in our case is 500, and it receives a bang to the left inlet, it will wait that number of milliseconds and then send a bang to its outlet. When the trigger sends a bang to the delay it will wait 500 milliseconds, and then send a bang to the 0 message, which will send 0 to the *~ object.

If you think this through, you can construct the reasoning behind this patch. Here's what we want to happen:

1. The message box "60" is pressed.
2. The note C will play.
3. After 500 milliseconds the note will stop.

To make that happen, when the message is pressed, telling the osc~ to play the MIDI note, we use a trigger to pass that along to the osc~ via the mtof converter. We also turn up the volume to full by passing 1 to the *~. Then we trigger a delay to count down 500ms and turn the volume off by passing 0 to *~.

With this deceptively simple patch, we've now gained some significant power over event timing in Pd. Bask in the glow of success for a little bit, but let's not stop there. Add six more messages with the values 58, 56, 58, 60, 60, 60 and connect them each to the left inlet of the trigger. Your patch should look like Figure 4, *Mary Had a Little Lamb*, on page 25.

Now press the messages in sequence from left to right. If you set up your patch correctly you should hear the notes to "Mary Had a Little Lamb." Enjoy!

Things to Think About

When the delay turns the multiplier to 0 and the sound stops, do you notice a click or pop from your speakers? Can you reason about why that happens?

Exercise: Using a trigger and delays, make a patch that plays the MIDI notes 60, 58, and 56 one second apart for 500ms. Play around with it; there's probably more than one way to do it.

Next Up

In this chapter we covered a lot. We created our first Pd patch and discussed objects and how to create them. We saw the difference between scalar values

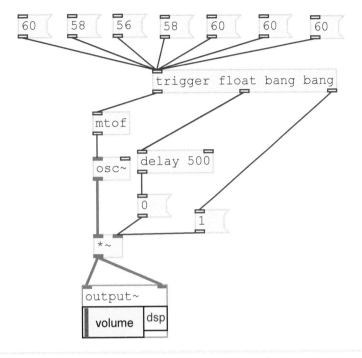

Figure 4—Mary Had a Little Lamb

and signals, and used a signal to make our first sounds with Pd. Then we covered a few ways of controlling a signal's volume and talked about messages and how they could work together with an oscillator to play musical notes.

Next we'll build some useful tools that will help us visualize the sound we're generating and control how sound happens over time. These tools will be integral to patches we build in the chapters after that.

CHAPTER 3

Building Controls

In this chapter we'll spend some time building tools and abstractions that will be useful in the more complex and interesting patches we'll build in the next chapter. This is an important step in the process of learning Pd—you'll be learning a lot about how patches function, but also building up a library of Pd components that will be useful to any type of patch you want to create in the future.

We'll start by building a patch that will allow us to get a *visualization*, specifically a live graph, of an oscillator's signal. Then we'll abstract that patch into a reusable control to get an idea of how Pd deals with code reuse and abstraction. You'll be able to drop it into any patch and get a quick idea of how a signal looks. Next, we'll build a reusable control called a *low-frequency oscillator*, or LFO, to control different aspects of sound in a periodic, repeating fashion. LFOs are common tools in synthesis, and we'll use one in the next chapter as we build some more interesting effects. Finally we'll build a tool called an *envelope*, which we will use in almost every patch in the rest of the book, and which you'll probably use in a lot of your patches, too.

When you are done with this chapter, you will

- Be able to visualize a signal
- Understand abstraction with subpatches
- Control periodic variations in a sound with an LFO
- Control a sound's volume over time with an ADSR envelope

Let's get started by visualizing the sounds we're making.

Visualizing Sound

When working with sounds it's at least as helpful to see what it looks like as to hear it. In fact, I recommend both listening to and looking at as many visualizations of sound as you can. Each sense will reinforce and support the other as you develop your mental vocabulary of the attributes and characteristics of sounds. Visualizations of sound are important because when working with sound it's possible to hear something only over time. That may sound trite or even tautological, but it's important: audibly experiencing a sound happens only over time. In contrast, you can look at a graph of a sound and quickly guess how it might sound without listening to it.

We've spent a whole chapter making and hearing some simple sounds; now let's get a look at them. Let's start by seeing what a 440 Hz sine wave looks like by creating a live graph of the signal output of an osc~. Make a new patch and create an osc~ 440. Next, create an object called tabwrite~ array1, and connect the osc~ to it. A tabwrite~ writes data from a signal into an *array*. Arrays are objects that can hold a table and a *canvas* to draw the graph of the array on. The easiest way to create one is to go to the Put menu and choose Array.

When you choose Array from the menu you'll be presented with a property-editor form for the array. Keep the array's name as array1, but uncheck Save Contents, because we don't want the data in the array saved with the patch. Keep the other defaults, and click OK. You'll end up with a large box split horizontally by a line. This is a graph of a new array named array1.

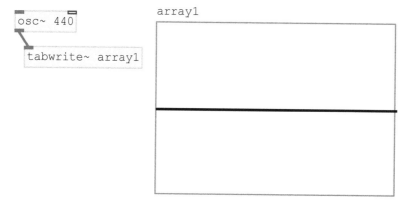

Now that we have an array called array1, the tabwrite~ array1 object can write signal data into it and it will be graphed. The only issue is that tabwrite~ doesn't automatically write data into the array, so we need to prompt it to do so. If we send a bang message to tabwrite~ it will fill up the array with as much data as it can hold, which will be supplied from the osc~. This is a common pattern with Pd objects: they react to the type of data they're sent. In this case the signal data from the osc~ is always flowing into tabwrite~, but it only writes data into an array when it receives a bang message.

That will give us only a snapshot of the signal, so we want to send it a lot of bangs to get a live picture of the signal. That's exactly what the metro object is for. It's short for *metronome*, and it will send out bang messages according to its float argument or right inlet, in terms of milliseconds. So a metro 100 will send a bang to its outlet every 100 milliseconds. Add one of these to the patch connected to the tabwrite~. A metro won't start automatically, however, so we need to send it a bang as well. Create an object called loadbang and connect it to the metro. A loadbang, just as it sounds, sends a bang message as soon as the patch is loaded. Now if you save the patch pictured in the following figure and reopen it, you'll see the graph of the osc~ 440 in the array labeled array1.

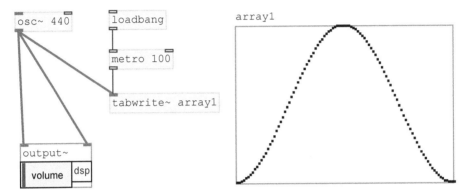

Now you're able to both hear and visualize a sound. Try changing the oscillator's frequency up and down, and see how the graph changes. Next let's look at turning this patch into a more flexible component for graphing a signal in whatever patch you need, by creating a *subpatch.*

Creating a Subpatch

A subpatch is a patch that works as a reusable component. It is stored as a separate file and can be loaded into another patch, called the parent patch. It can have inlets and outlets just like the other objects we've worked with so far. It can also draw part of its canvas on its parent patch, which is great for exposing user-interface elements to control the subpatch. We'll turn the simple graph we just made into a subpatch, so you'll be able to add it to any patch quickly to view a signal.

Create a new patch and save it to a directory of your choice. Call the patch grapher~, which will create a file called grapher~.pd. Note that we used the suffix ~ to indicate that this subpatch takes a signal input. The suffix is not strictly necessary, but adding it will show how to follow the convention of using ~ when working with subpatches.

Select the loadbang, metro, tabwrite~, and array from our previous patch and cut them just as you would cut text from a document. On the new, empty canvas, paste those objects. Arrange the rest of the objects around the array with a bit of space at the top for a new object. Next to the top left of the graph create an object and type inlet~ signal and connect it to the tabwrite~. Now you should have a canvas that looks like this figure.

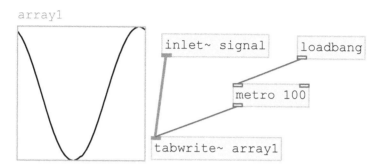

An inlet~ creates an inlet like we've seen at the top of other objects: one of the little shaded boxes to which we can drag a connection. It also takes an argument, which is whatever you want to name the inlet—the argument has no effect on the workings of the subpatch; it's more like a comment to help the subpatch developer call out what he had in mind when designing the subpatch.

Using a Subpatch

Now we'll use the subpatch as a component of another patch. Make sure grapher~.pd is saved, and go back to the patch we cut the objects from. Save this patch as A440_graphed.pd. Now, to the left of the osc~ create a new object called grapher~ as shown here.

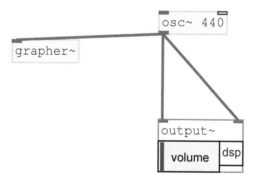

Note that both patches are in the same directory and the filename of the subpatch we created was also grapher~.pd. You may have to close the patch and reopen it to make Pd reload the subpatch with its new inlet~. This is a simple example of how subpatches work: an object can refer to a filename from the local directory and Pd will load that patch and work with any inlet, inlet~, outlet, and outlet~ objects defined within.

The problem we have now, though, is that we want our subpatch to graph on the parent patch the signal going to its inlet, but it doesn't do that yet. To make it work we'll need to go back and open grapher~.pd—alternatively, in Run mode we can click a subpatch and it will open in a separate window.

When the subpatch is open, right-click anywhere on the canvas holding all the objects in the subpatch and choose Properties. Make sure Graph on Parent and Hide Object Name and Arguments are both selected. The first property tells Pd that this subpatch wants to draw to the screen of its parent patch. The second property tells Pd that it should not print the subpatch's name—that is, grapher~—to the parent patch's screen. Click OK, and you should return to the subpatch. Notice that there is now a red box drawn on the canvas. This red box represents the area that Pd will draw on the parent patch. If you right-click and choose Properties again, change the margins from 100 to 0, and click OK, you'll notice that the red box jumps to the top left of the patch. Now select everything and move it so that the array1 graph's top-left corner fits inside the red box's top-left corner. Now edit the canvas's properties again and resize the red box to fit around the graph. Save the subpatch, close it, and return to the A440 patch. Instead of an object called grapher~ you should see something like the patch shown here.

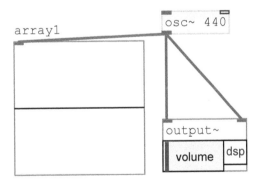

Because we edited the subpatch to Graph on Parent, Pd will draw the red box's contents on this parent patch, and because we chose Hide Object Name and Arguments, you will not see grapher~, just the graph. All you should see in the graph, however, is a flat line. To make the graph work, turn DSP on; you should see a smooth sine wave graphed as in the following figure. Why does DSP need to be on for the grapher to work? Remember that DSP means *digital signal*

processing; in other words, this toggle says whether Pd should process signals or not, not just whether it should send sound to the sound card.

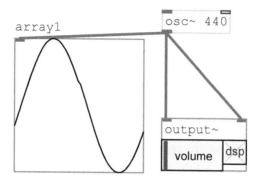

Now, using this subpatch we can quickly and easily connect a graph to any signal in a patch and visualize it. Before we leave the subject of subpatches, there's one more topic we need to discuss: local variables.

Working with Local Variables

The grapher~ subpatch as it's currently designed will not work if we want to use it more than once in a patch. To illustrate the problem, let's create a patch that graphs two oscillators. Create a new patch, save it next to grapher~.pd, and call it double_graph.pd. Create an osc~ 440 and a grapher~ and connect osc~ 440 to the grapher. If DSP is on, this should graph the 440 Hz sine wave as before. Now create another object: osc~ 880. Your patch should look like the following image.

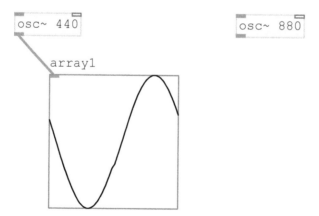

Now create another grapher~ below the 880 Hz oscillator. If you connect the osc~ to the new grapher, it doesn't work. Now switch back to the main Pd window, and you should see the logs screaming at you.

```
warning: array1: multiply defined
warning: array1: multiply defined
warning: array1: multiply defined
warning: array1: multiply defined
warning: array1: multiply defined
warning: array1: multiply defined
warning: array1: multiply defined
warning: array1: multiply defined
```

Delete the second grapher~ to stop these warnings. This illustrates an important point about arrays and naming in general in Pd. The problem here is that we have a subpatch with an array named array1. The name array1 is in the global namespace. It's defined for the whole Pd instance, and if you try to define another array with the same name, Pd will not know which one to address and will start to complain loudly in the logs.

To fix this problem and make subpatches with things like arrays viable, we can use a local variable. Open grapher~.pd, and you'll notice that even now Pd complains about array1 being defined multiple times. Turn off DSP, and the warnings should stop. Now right-click the graph for array1 and change the name of the array to $0-array1. Also change the tabwrite~ argument to $0-array1.

$0 is a *dollar-sign variable* (it is prefixed with a dollar sign). Pd replaces the $0 with a unique number for each instance it finds in any patches it loads. You may have run across a similar concept in other programming languages. Now within the grapher~ subpatch we can refer to $0-array anywhere we like, safe in the knowledge that we mean the local array1. Save grapher~.pd and return to double_graph.pd. Now create a second grapher~ and connect osc~ 880 to it, and you should see both waves graphed.

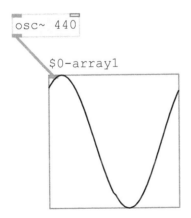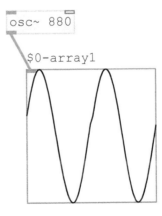

With the ability to visualize two signals at once using the grapher~ subpatch, you can see a nice visual confirmation of the fact that the oscillator at 880

Hz is double the frequency of the one at 440 Hz. Now let's look at building a few different subpatches that will be useful as controls later to help us introduce some motion to the sounds we create.

Making Sound Move

So far the sounds we've made have mainly acted as a support to learning how to work with Pd, but they haven't been interesting sounds per se. One component of real-world sound we haven't yet reproduced is movement. Real sounds usually change over time. Whether it's the gentle undulation of waves on a beach or the wind in the trees, or the dynamics of an impact like two wineglasses clinking together, there's usually a component of change over time that a sound designer needs to understand and be able to reproduce.

We'll build two components that produce change over time two different ways. The first is a slight variation on the oscillator we've already been using to make sound, called a low-frequency oscillator or *LFO*. The second is a way to control the volume of the sound in stages, called an *envelope*. First, let's talk about an important but subtle difference in how we'll use signals to make the sound move and change over time.

Controlling a Signal with a Signal

As we saw in the section on visualization, an osc~ produces a wave that undulates between two values, usually 1 and −1, over time. When that wave is used as a signal and sent to Pd's output, the wave moves a speaker and makes sound. If we instead used the signal from an oscillator to change some property of another signal, perhaps the amplitude or frequency, then the first signal, the one doing the changing, is said to be in the *control domain* while the second signal, the one making sound, is said to be in the *signal domain*. Put simply, signals can be used to make sound or to change other signals. The process of using a signal to change another signal is called *modulation*.

The modulation's speed is significant in an interesting way. If we take a simple signal like a sine wave at 440 Hz and slowly turn the amplitude from 0 up to 1 and back down to 0 over 10 seconds, you'll hear a sound fading slowly in and out. If we start to turn it up and down faster and faster the sound that you hear won't simply fade in and out, but will start to become a new, more interesting sound. This process is called *amplitude modulation*, or AM for short, and it's a way to use fairly simple components, like oscillators, to create complex sounds.

A similar technique is *frequency modulation* (FM), which is where a control signal modulates another signal's frequency. FM can produce even more complex sounds than AM. Let's try out both of these techniques and see a small range of the possibilities. To do this we'll create a tool called an LFO, which is a useful tool to have, and one we'll make use of in later chapters.

Designing an LFO

A low-frequency oscillator is a common control mechanism. Just like the oscillators we've already used with osc~, it creates a signal. The "low frequency" part means that it oscillates below audible frequencies. Generally speaking, the range of audible frequencies for humans is 20 Hz up to 20 KHz. In practice 20,000 Hz is higher than most people can hear because we lose high-frequency perception as we get older or as a result of hearing damage from loud sounds or just plain aging. Our low-frequency perception generally stays at a lower limit of around 20 Hz.

An LFO's function is not to produce low-frequency sounds, but rather to control how another signal acts over time. An oscillator with a period in the seconds to tens-of-seconds range can help produce certain sorts of sonic phenomena. We'll build an LFO that we can load as a subpatch, which will have controls for the frequency and the amount, or depth, of the effect, and then two different patches to test out its effect on our old friend osc~ 440.

Let's look at the specs for our LFO control:

- It should have a way to adjust frequency between 0.05 Hz and 200 Hz. A lower end of 0.05 Hz will let our LFO have a lower period of 20 seconds. Remember, hertz is a measurement of frequency, which is the inverse of the period in seconds. 200 Hz on the higher end is pretty high for an LFO, but it will allow us to do a little playing around with amplitude and frequency modulation and learn a little more about how they sound.

- It should have a way to adjust the depth of the effect between 0% and 100%. We want to be able to adjust how much effect the LFO has, between full effect and no effect.

- It should have a signal outlet. The LFO should output a signal, so we can use it to affect other signals.

Now let's talk about building a patch to meet these specs.

Building an LFO

As we get into more complex patches it will take more space to walk through the individual steps to create a patch, so from now on we'll rely more heavily on looking at a patch to understand how it's made. One of the great things about Pd is that looking at a patch tells you almost everything you need to do to reproduce it.

Create a new patch called lfo~.pd and save it in the same directory as grapher~.pd. Then re-create in the patch what you see in the following figure.

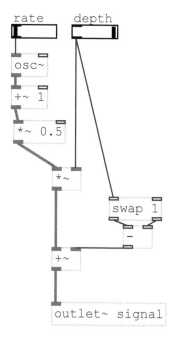

This LFO is meant to be used as a subpatch, like grapher~. This diagram shows the patch with most of the controls in place, so you can see them a little more easily. Then we'll go back to the parent patch and add other labels and controls that we want the user to see. That will get a little cramped visually. Let's go over the patch so far.

Controls and Display

First we have rate and depth controls. Those controls are horizontal sliders with a width of 30 and a height of 10. The rate slider's *output-range* properties are set to 0.05 for left and 200 for right, which will give us the Hz range we want. The depth slider's *output-range* is 0 to 1 for 0% depth to 100% depth. The Rate and Depth labels on the sliders can be added from their property windows.

Getting the Rate in Range

The rate slider is also connected to an osc~ object with no argument, so it controls the oscillator's frequency. This is the LFO's engine, so to speak.

Signals in Pd generally have values between 1 and –1, but we want the LFO to output only a positive number. To make this happen we do a little massaging of the signal coming out of the osc~ so that it never goes below 0—we add 1 and then halve the value so the lowest possible signal value is 0 and the highest is 1. The signal is then sent to a *~ to mix in the depth control's effect.

Getting Depth Working as Expected

The depth value is sent to the *~ object to regulate the osc~ signal, but then we have to do a little more work to make sure the depth works the way we'd expect. If the depth is set to 0, we don't want the LFO to output a signal of 0. When the depth is set to 0 the intention is that the LFO has no effect when, for instance, we multiply the signal from an osc~ with the LFO. To make this work we need a depth of 0 to output a signal of 1. We send the output of the depth control to a swap object with an argument of 1. This object will send the scalar value received at its left inlet to its right outlet, and either the value of its right inlet or its argument to the left outlet. These values are sent to a subtraction object (-), which will send the value of 1 minus the depth control's value to its outlet. That way a depth of 0 will cause a 1 to be calculated at this stage, and a depth of 1 will calculate a 0.

Mixing Rate and Depth for Output

The value of *1 - depth* is sent to a +~ object to be added to the signal derived by multiplying the rate by the osc~. Therefore if the depth is 1, it will not affect the already fully affected osc~ output, but if it's 0 it will add 1 to the signal of 0 coming from *~ and send a signal of 1 to the outlet~.

Adding Readout Controls

Now let's go back to the top of the patch and add some controls that will calculate and display information to the subpatch's user—useful controls that will show when the patch is used in a parent patch. Just like with the grapher~ patch, right-click on the canvas and choose Properties and set them as follows: Graph-on-Parent checked, an X size of 100, Y of 85, and margins of 0. Figure 5, *Final LFO Patch*, on page 38 shows the patch again, this time annotated; you can create comments with the key sequence ^5. Also, note that in this figure I added two inlet objects, one each for rate and depth, to allow the parent patch to specify the rate and depth.

When a patch is set up to graph on its parent, remember that the red box shows the area that will display on the parent. Inside this box is where the rate and depth controls are arranged.

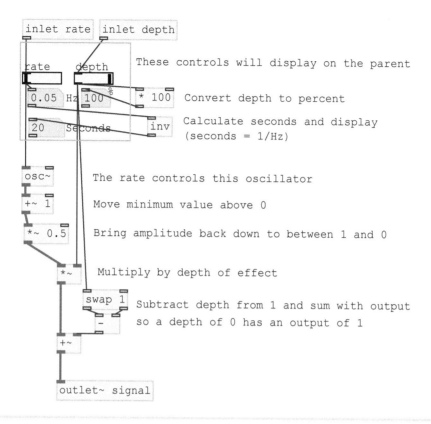

Figure 5—Final LFO Patch

To provide a bit more information to the LFO's user, there are boxes underneath. Numbers are created with ^3. A number can display a scalar value sent to its inlet. There is a number for the rate slider's direct output, which is labeled Hz, and another underneath labeled Seconds. We can edit the label in the number's properties.

Hertz is defined as a rate per second, so to convert Hz to seconds we can simply invert it. To derive seconds for the rate control's Hz value, the Rate slider is connected to an inv object, which inverts the scalar value and sends it to its outlet, which is then connected to the number displaying seconds. The Depth slider is connected to a *100 object, which converts its scalar value to a percent, and sends that display to the number labeled with a percent symbol (%).

Now that we have an LFO control that will allow us to easily control the rate and depth of a control signal, let's test it out.

Modulating Amplitude with an LFO

The first test we'll make will be to vary the signal's amplitude from an osc~. When used musically, this is called *tremolo*. Create a new patch and save it in the same directory as lfo~.pd, and name the file lfo_tremolo.pd. Create an osc~ 440, connect it to a *~, and connect the signal-multiplication object to both sides of an output~. Next, create an object called lfo~ and connect it to the other side of the multiplier, and you should have a patch that looks like the figure here.

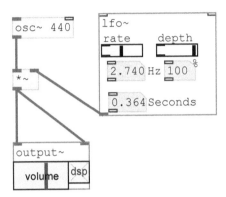

Turn up the volume on the output~, and you should hear the 440 Hz tone slowly go up and down in volume. Try a few experiments:

- Move the rate slider back and forth slightly, keeping it below 10 Hz. You can move a slider with extra precision by holding down Shift and then moving the slider's handle.

- Move the depth slider back and forth. Notice that at 0% depth the 440 Hz tone is constant, which proves what we achieved with the swap 1 object inside the LFO subpatch.

- Setting the depth slider at 100%, move the rate slider slowly past 20 Hz. Notice how at around 40 Hz the effect starts to change from simply increasing and decreasing the volume of the osc~; instead some other tones start to appear.

This is a simple example of adding an element of motion to a sound. The LFO's nature is such that the motion it produces is regular. Taking the LFO into audible frequency range also illustrates how some interesting sonic content can come from unexpected phenomena. This is an example of *amplitude modulation*. We won't use it in our examples, but it's an interesting process to experience.

Modulating Frequency with an LFO

Now let's do something slightly different with the LFO and modulate an osc~'s frequency instead of its amplitude. When used musically, this effect is called *vibrato*. Create a new patch, save it in the same directory as lfo~.pd, and call it lfo_vibrato.pd.

Instead of passing an argument to the oscillator, let's create an osc~ and connect it to both sides of an output~. Now create a *~ object *above* the osc~ and connect it to the osc~. Then make a number above the signal multiplier and connect that to the left inlet. This is so we can vary the oscillator's frequency. Connect an lfo~ to the right inlet of the *~, and the patch will look like the following figure.

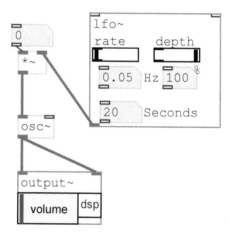

Click and drag upward on the number object, setting it to a medium-high audible frequency, like our old friend 440 Hz. You'll hear a slowly rising and falling tone. This example is similar in construction to the amplitude-modulation patch, but with strikingly different effects. Frequency modulation is a powerful technique that we'll look into more later. The patch is more of an experiment with the LFO than a practical FM example, but it does give you a chance to see how modulating frequency affects sounds.

This patch is the most complex we've built so far, so nice work! This component will be a great one to have, and we'll put it to good use in an example in the next chapter. For now, though, let's move from the LFO to another tool that adds motion to sounds, an envelope.

Things to Try

As always, playing around with the controls is a great way to learn. Try a few things:

Set the depth to around 10% and the rate to around 4 Hz. This is a fairly extreme but recognizable vibrato tone.

Sweep the number value between roughly 220 and 440. Sounds sort of like the "mad-scientist lab" sound from old movies, doesn't it?

Leave the number set to around 440, but increase the depth to 100%. Sounds a bit like a sci-fi machine.

Raise the rate slowly to maximum and notice how the tone is completely changed until finally it's a hollow-sounding tone very different from the smooth sine wave at 440 we've been using.

Building an Envelope

Whereas the LFO is a tool for creating repeating motion in a sound, an envelope is for creating single motion events. The name "envelope" comes from the analogy of opening and closing a paper envelope, so an envelope is generally for increasing some parameter and then at a later time decreasing it. A common use is to increase and decrease the volume of a signal over time. We'll use this envelope in almost every patch we make from the next chapter until the end of this book, and you'll find this to be a valuable tool in your own patches.

Designing an Envelope

We'll build a subpatch that matches the specs of a classic *ADSR* envelope. That stands for *attack, decay, sustain, and release.* Envelopes can have more or fewer parameters than these four, but the ADSR is a time-honored formula that may be familiar to you if you've worked with musical synthesizers. The envelope's parameters are described here:

- Attack — The attack portion of the envelope is when it raises from zero to a given value, the *attack level*, over a given period of time.

- Decay — After the attack time is complete and the attack level is reached, the decay time specifies how long the envelope should take to reach the sustain level.

- Sustain — The sustain level is where the envelope stays until it is triggered to stop.

- Release — When the envelope is triggered off, the release time specifies how long it takes to move from the sustain level to zero.

A graph of these parameters may help in making sense of the idea.

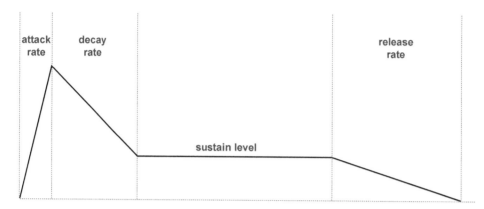

The attack time, controlled by the attack rate, takes the envelope to the attack level. The decay time, controlled by the decay rate, takes the envelope to the sustain level. Then after a trigger to release, the envelope takes the release time, controlled by the release rate, to return to zero. Of course, any of these parameters could be zero—for instance, the sustain level. If the sustain level were zero then the envelope would effectively shut off automatically after the decay time was complete.

Creating the Envelope Patch

The envelope patch we'll make is fairly complex. Reproduce the subpatch from Figure 6, *The Envelope Patch*, on page 43, save it in the same directory as lfo~.pd and the other patches we've worked with so far, and name it adsr~.pd.

Now let's talk through what's going on here. As you look at the patch, you'll notice a number of new things. First of all, this patch has six objects we haven't seen yet. You also may not recognize some of the objects we have used before, because they're abbreviated. Finally, there are some dollar-sign arguments greater than 0. Let's go through these objects one by one, and then break down the subpatch to understand what it does.

Abbreviating Objects

Abbreviation is a new concept here. Instead of typing out an object's whole name, Pd lets you abbreviate the name to save space. For instance, in the subpatch, you'll notice that the flow goes from the inlet labeled Trig into an object called sel and then through its right outlet to an object called t. We'll

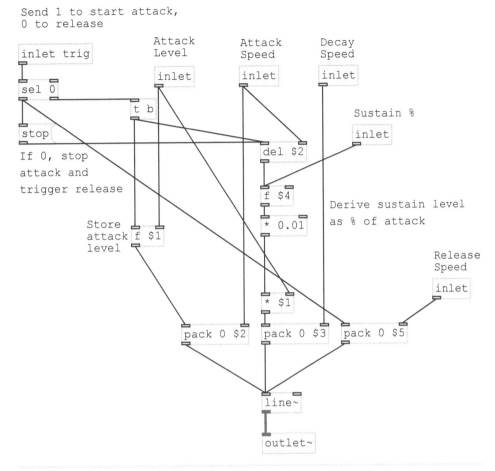

Figure 6—The Envelope Patch

look at sel in a second because it's new to us, but the t b object is actually a trigger with one argument, bang. Pd recognizes the t as an abbreviation of trigger and b as an abbreviation of bang. The other object abbreviated in the subpatch is delay, which is abbreviated del. Not all objects have a valid abbreviation—using the context help on the object will tell you if it has one, but abbreviations do save space and typing.

More Dollar-Sign Variables

The next new concept, which builds on the $0 we saw from the section on local variables, is dollar-sign variables corresponding to arguments. If a subpatch takes creation arguments, for instance adsr 1 10 500 10 10, then each of those arguments will be stored in order inside numbered $ arguments.

Joe asks:

Why Would We Want Both Inlets and Arguments in a Subpatch?

Adding inlets and arguments, which fill in dollar-sign variables, to a subpatch gives us flexibility and an easier way to set defaults. If your subpatch has the ability to take both arguments and inlets, it's easy to configure it with arguments if you never need the parameters to change. Or, if you do want some values to change and others to stay as defaults you've set in arguments, you can choose which ones to control from other atoms.

By the way, if you set a subpatch's arguments and control the same values with inlets, the inlets will change the values as expected. The arguments are set when the subpatch is loaded.

In adsr 1 10 500 10 10 there are five creation arguments, and so in the subpatch there will be available $1, $2, $3, $4, and $5 containing those values, from left to right. You may be familiar with this concept from shell scripting or languages like Ruby and Perl that use the same conventions.

New Objects

Now let's go through the new objects one at a time, starting at the bottom:

- *line~* is an object that outputs a ramp from a stored value to another number, based on messages you send it. For instance, line~ defaults to 0. If you sent it a message of 1, it would jump up to 1 and send that value to its outlet. It's possible to send a list of numbers in Pd by separating the elements of the list with a space in a message. If line defaults to 0 and receives a list from a message *1 500*, then line would take 500 milliseconds to reach 1 and then have 1 stored as its new value. While it was changing from 0 to 1 it would send the interpolated values to its outlet.

- *pack* is an object that will take individual values sent to its inlets and pack them together into a list. If the left inlet receives a 1 and the right a 500, then its outlet would be a list containing 1 and 500.

- *float*, abbreviated f, is an object that simply stores a value, either initialized with an argument or sent to its left inlet.

- *select*, abbreviated sel, when given a value to its left input, will output that value to its left outlet if it matches its argument, and if it does not, it will output the value to its right outlet.

- *stop*, as seen in the message box at the top of the subpatch, is not an object, but rather a special kind of message: the opposite of bang.

That's a number of new objects. Armed with that knowledge, let's look at the subpatch from the bottom up.

The Envelope Patch from the Bottom Up

The line object makes the smooth transition from one stage of the envelope to the next. It's controlled by list messages from the array of pack objects above it, which each correspond to a stage of the envelope, from left to right: attack, decay, and sustain.

The leftmost attack pack gets its list from the f object that stores to the subpatch either the first argument, $1, or the value from the second inlet, level. The pack's right inlet comes from either the second argument, $2, or the third inlet, attack. With the pack thus initialized, when the first inlet, trig, receives a 1 it will trigger the float to send its value to the pack, which will cause the list to generate a smooth ramping output from 0 to the attack level, taking the attack speed in milliseconds to do so.

The middle pack object controls the drop from the attack level down to the sustain level, taking the decay speed in milliseconds to get there. This happens because when the ADSR is triggered, it starts the delay object, which counts down until the attack speed is complete, and then triggers the f containing the sustain level, expressed as a percentage of attack level, to output its value through a * 0.01. This multiplication object converts from percent and sends its value into another * object, which has the attack level set through its right inlet. This puts the converted sustain-level value into the left side of the pack object, which then sends the sustain value and decay time in milliseconds to the line.

At this point the ADSR will remain constant at the sustain level until it receives a 0 to its leftmost inlet. When the 0 is received the third pack will send a list containing 0 and the release value in milliseconds to the line, which will ramp down smoothly to 0.

This is a pretty advanced patch compared to what we've done so far, and again, it will be a great tool to use in more interesting patches later. If it's not clear what it does yet, testing it out will be the best way to understand it.

Testing the Envelope

Create a new patch, save it in the same directory as adsr~.pd, and call it adsr_test.pd. This will be a pretty simple patch; create an osc~ 440 connected to a *~ connected to an output~. Then create two messages: 1 and 0 connected to the left inlet of an adsr 1 10 500 10 10 object. Connect that to the right inlet of the *~. This should look like the following figure.

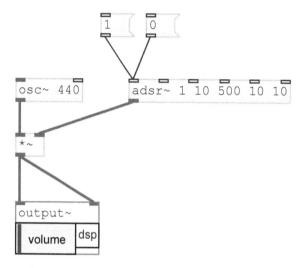

Now if you press the 1 message it will play the 440 Hz tone, first loudly and then continuously at a lower volume. If you then press 0, the tone will fade out quickly. Once again, the arguments in order are as follows:

- Attack level — How loud the attack should be (assuming the ADSR is controlling an audible signal)

- Attack time — How long the envelope should take to reach the attack level

- Decay time — How long the envelope should take to decay to the sustain level

- Sustain level — The percentage of the attack level at which the envelope should hold after the attack is over

- Release time — How long the envelope should take to reach zero after receiving a message of zero

If anything is unclear, follow the flow of the connections from each inlet, perhaps with some test values, and think about what happens at each stage. An envelope is a very useful tool to have, and we'll use it very soon to make some realistic, moving sounds.

Things to Try

As always, playing around with the controls is a great way to learn. Try a few things:

- Make the attack time long, like 10 seconds.
- Make the release time long, like 20 seconds.
- Make the sustain 0, effectively making the envelope an AD, or attack-decay, envelope.

Next Up

In this chapter we laid a lot of ground work for the coming chapters. We covered arrays and how to graph a signal, and then created a useful, reusable component—a subpatch. Then we loaded the subpatch into a parent patch, discovering how Pd's code reuse works. We also learned about how to keep variables local to a subpatch.

We built some useful tools: an LFO to control periodic motion in a sound, and an envelope to control single motion events. Along the way we ran into a number of new objects and features of Pd.

In the next chapter, we'll put the controls we created in this chapter to use as we move away from our standard osc 440 patches and create some effects!

CHAPTER 4

Creating Effects: Real-World Sounds

It's time to have some fun. In this chapter we'll start to apply what we know about Pd and use the tools we've built to create some sound effects. The sounds are relatively simple to create, but are surprisingly realistic for the amount of work it takes to create them.

We'll start with the simplest patch, simulating waves with white noise, a filter, and a low-frequency oscillator (LFO). Then we'll build up a bit with a patch simulating wind with noise in both the signal and control domains, and an envelope for more control. Finally, we'll do our first more technical sound analysis by looking at a recording of a wineglass being tapped, and then reproducing the sound it makes.

When you are done with this chapter, you will

- Understand what noise is, and how it can be used
- Know how to shape sound the way you want
- Know the principles behind two types of synthesis: subtractive and additive
- Have a basic understanding of sound analysis

Each section in this chapter follows the pattern of a general description of the sound we want to make, an analysis of what's going on in the sound in the real world, the approach we'll take in Pd to reproduce the sound, and a discussion of the patch. From now on we'll start with completed patches or subpatches and break them down instead of talking through building the patches step by step.

Let's get started by making a simple patch that reproduces the sound of waves at a beach.

Waves

The sound of waves at a beach is a nice, calming sound, and is a great one for our first shot at designing environmental ambience. If you've ever been to the beach, you can probably recall the waves as an undulating, mellow, hissing sound that repeats over and over. It depends a bit on the makeup of the beach, but the common element is the water.

I remember visiting a beach, French Beach, in British Columbia. The beach there isn't sand; it's made up of fist-sized rocks, and as the waves rolled in and out they made this wonderful *tocking* sound multiplied countless times by the sheer number of the rocks that added an uncommon and interesting element to the waves. I'd love to be able to capture and analyze that sound. For our purposes, though, let's consider a sandy beach—it's familiar and relatively easy to approximate with some simple tools in Pd, so it's a great place to start practicing sound design. Let's begin with a simple analysis.

Analysis

The first thing to note about waves is their periodic nature. That's one of the reasons they have such a lulling effect on you as you sit on the beach—the regularity of the sound of the water building up forward motion, reaching up the beach, and then flowing back down the sand into the body of water over and over again.

The second thing to note is the materials involved. The effect of water flowing around and through itself and over and around the material of the beach is a complex physics problem to describe, involving hydrodynamics and a whole set of interesting theorems. Our model doesn't have to take that deep of an approach, though, because the sonic effect of all this activity can be simulated with *noise*. In sound production, noise has a more specific definition than we use in day-to-day speech; it means some sort of random signal.

Noise is classified into different *colors*, drawing an analogy between sound and light and the distribution of energy across the visible spectrum. *White noise* is a random signal with a uniform distribution across all frequencies. For comparison, there is also *pink noise*, with the frequencies weighted toward the lower end, and *blue noise*, with the frequencies weighted toward the higher end of the spectrum. Because of the way our ears perceive sound, white noise sounds higher-pitched than you might expect from the even frequency distribution. Most times you hear *noise* in this technical sense you're probably hearing about white noise. This is the case in Pd, so we'll stick with "noise" instead of "white noise."

Noise is close to the *spectrum* of water flowing in and around sand while the grains of sand raise and settle, bumping into each other. The spectrum of a signal is the distribution and strength of the frequencies in the signal, and reproducing the spectrum of a sound is the greater part of reproducing the sound.

Approach

We'll model the sound of waves on a beach by doing the following:

- Using an LFO to model the timing
- Using a noise generator to model the spectrum

Pd supplies us with an easy way to make white noise. Unprocessed white noise sounds unnatural, so we'll tune it to our taste using a filter, which removes frequencies from a signal. Since we don't have any more specific requirements for reproducing the sound of waves, we're free to apply a filter and then adjust it until it sounds right.

The Patch

Create a new patch named waves.pd and save it. Reproduce the patch shown in this figure.

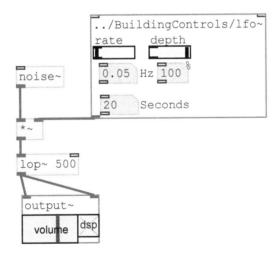

First, note that the sample code has this patch in a sibling directory to the directory containing the patches from the last chapter, including the lfo~.pd subpatch. Pd will load subpatches from relative paths, and in this case the path to the subpatch is ../BuildingControls/lfo~.pd. You can either use the relative path to the LFO subpatch from your system or copy the subpatch into the same directory as this patch and refer to it as lfo~. The LFO is then connected to a *~ to regulate the amplitude of the noise.

The aptly named noise~ object connected to the other side of the *~ generates white noise. The lop~ between the *~ and the output~ is a filter to tune the noise a bit. "Lop" stands for *low-pass filter*, which is a filter that cuts off frequencies above a given frequency, which it takes as a numeric argument, and allows lower frequencies to pass through and remain in the signal. In the figure the lop~ has a relatively low cutoff frequency, but try different ones out and see what you think.

The period of the waves can be controlled with the rate of the LFO, but notice how rates of less than 10 seconds start to sound unnatural. The depth of the LFO is also subtly important to the effect; a depth of 100% sounds unnatural because waves rarely become completely silent. Try something between 75% and 80%.

Experiment and play around: imagine the sound of seagulls and boats in the background, or perhaps children laughing and playing, and you can see how amazing it is that some basic building blocks can produce realistic sounds. Next we'll explore using noise a bit more as we reproduce the sound of wind.

Wind

Wind, too, is a great ambient effect that's relatively easy to reproduce using synthesis. Let's consider what kind of wind we want to reproduce. To get the most out of the effect let's make it a fairly windy day in something like a grassy field with a few trees, so we'll hear a breeze with light gusts every now and again, but sometimes experience stronger, more steady gusts. We'll make it so that the stronger gusts are controllable, which could be useful to tie to some event taking place on a game screen, for instance.

Analysis

As with the waves patch, this analysis is intuitive rather than scientific. The method of reproducing wind is similar to waves in that white noise can reproduce the spectra created by the chaotic flow of air around materials of different shapes.

One big difference is that wind is nonperiodic, unlike waves on a beach. Wind blows harder, then softer, with gusts at unpredictable times. One last thing to note is that wind is a subtler sound than waves, so we'll want to use a more drastic filter to pick out the frequencies we want to be in the signal.

Approach

Let's consider the structure of the patch in both the signal and control domains.

Signal Domain

As with waves, the signal domain will be based on noise. To get the spectrum we want we'll use a bp~, Pd's *band-pass filter*. Whereas a *low-pass filter* allows only frequencies below a certain frequency to pass through, a *band-pass filter* allows only frequencies a certain distance *around* a given frequency, which are called the *passed band*. Band-pass filters usually have a parameter called Q, which controls the width of the passed bands in an inverse relationship, so higher Q values pass a narrow band and lower values pass a wider band.

Control Domain

We have two goals in the control domain for this patch:

- A steady, quiet wind with small, random gusts
- A way to control and produce louder gusts for a period of time

We'll control starting and stopping the wind and producing the louder gusts with an ADSR (attack, decay, sustain, and release) envelope. That part is much like the test patch we made last chapter. The small, random gusts are a little more tricky. We need some way to introduce randomness into the control domain.

Luckily, we don't have to look far, because that's exactly what noise is: a random signal. So we'll use noise~ objects in both the signal and control domains, one for the wind sound and one to produce gusts, which are just small amplitude jumps to the signal. We'll also use a series of filters to regulate how much effect the control noise has.

We'll split the patch into the wind-making mechanism and the controllable part by using a *subwindow*, or internal subpatch: a subpatch that's not stored in a separate file, but rather in the patch itself.

The Patch

First let's work on the main patch. Create a patch in the same directory as waves.pd and call it wind.pd. Then create an adsr~ either by copying the one we made to the same directory or by referring to it with a relative path. Connect the adsr~ to a *~ and both sides of the *~ to an output~. You should have something like in Figure 7, *The Main Patch for Wind*, on page 54.

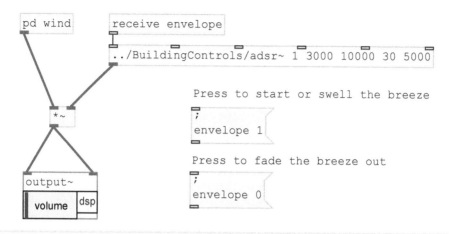

Figure 7—The Main Patch for Wind

The left side of the *~ has a new type of object connected to it, called pd wind. Create this now. Notice that once you click away from editing the object, Pd automatically opens a new window for you. This syntax creates a subwindow.

Subpatches and subwindows are both great ways to build some abstraction into your patches and focus on the various responsibilities of the different parts of a patch. In this way, Pd is like any other programming language. Subwindows are just like subpatches, but you create subwindows when the function you want to perform is specific to the current patch, and subpatches in separate files when you want to build a reusable component.

The wind subwindow looks like Figure 8, *The Wind Subwindow*, on page 55.

The left side of the window under the comment "Wind signal" starts off with a noise~ object connected to a bp~. This is the band-pass filter discussed earlier. The arguments 800 1 correspond to the center frequency of the filter, 800 Hz, and the Q of the filter, 1. This is a low Q value, which means the band is fairly wide. Here again, 800 Hz sounds fairly good as the center band, but it's not scientifically chosen. Play around with different frequencies and Q values to get an idea of how they sound.

The signal comes out of the bp~ and into a *~ so we can vary it with the control signal simulating random gusts. That signal chain also starts out with a noise~, but then we pass the signal through a series of very drastic lop~ filters set to 1 Hz. Filters don't lop the sound off (pun intended) directly at the filter frequency, but rather they have a curve to them, so using two filters in series like this makes the cut-off steeper than using one. These filters keep out

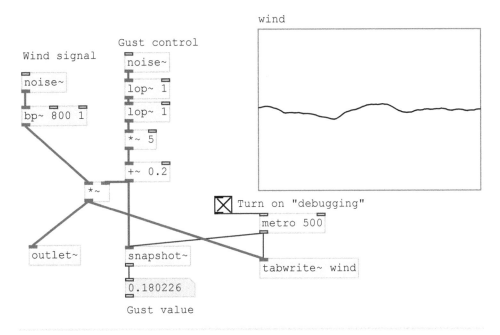

Figure 8—The Wind Subwindow

almost all of the signal from the noise~ object so that only a very subtle signal gets through. It's so subtle that we have to increase the magnitude by 5 with a *~ to get it in a range where the effect will be noticeable.

We don't ever want the control signal to reach 0, because that would stop the sound of the wind. To avoid that we add 0.2 to the control signal with a +~. The control signal is connected to the right side of the *~ to regulate the signal from the left side.

So that we can see what's going on, there is also a graph of an array called wind, a snapshot~ of the control signal fed to a number, and a metro 500 controlled by a check box, which you can turn on to debug the patch. When the patch is running, watch the gust value and notice the small amount of amplitude regulation.

The randomly regulated signal is sent through the outlet~ and back to the main patch. There, the message boxes containing envelope 1 and envelope 0 require a little explanation. This syntax where the message contains a semi-colon (;) followed by a new line is a way of broadcasting a named value. To understand this better, consider the receive envelope object connected to the adsr~. When you press the message containing envelope 1, Pd sets a global variable named envelope to the value 1, and any receive objects with an argument envelope will be triggered, sending the value of envelope to their outlets. The same goes for the message containing envelope 0.

The 1 and 0 values sent to the receive connected to the adsr~ serve to trigger and stop the envelope, which starts and stops the wind. Note the arguments to the adsr~, in order:

- 1—An attack value of 1 means that the attack portion of the envelope will reach 1.

- 3000—This attack rate in milliseconds means the attack portion of the envelope will take 3 seconds.

- 10000—The envelope decay will take 10 seconds.

- 30—The sustain value will be 30% of the attack value, or 0.3.

- 5000—The envelope will stay at 0.3 indefinitely until it receives a 0, when it will take 5 seconds to reach 0.

And that's our wind patch! Try it out; click the envelope 1 message and listen to how the wind swells and then drops. Listen to how there are small gusts blowing through randomly. Then when you want a larger sustained gust press the envelope 1 message again. Remember that this could be controlled inside of a game to coincide with wind control that also affects what the player sees on the screen. When you want the wind to fade out, press the envelope 0 message.

Next we'll take a little bit more time to analyze a real-world sound and reproduce it. But before we do, let's take a moment to talk about a little theory behind what we're doing here.

A Short Interlude: Synthesis Types

Before we go on, let's talk a bit about what we've been doing in this chapter at a higher level. Synthesis, you've learned, is the process of putting together sounds from component parts. There are different methods of synthesis, and so far in this chapter we've used one: *subtractive synthesis*. As the name implies, subtractive synthesis is the process of taking away frequencies to leave behind the desired ones.

We used subtractive synthesis to tune the white noise to the frequencies that sounded best in the waves and wind patches. We did this with two kinds of filters, a low-pass filter and a band-pass filter. Here is a list of filter types:

- Low-pass filters remove high frequencies.
- High-pass filters remove low frequencies.
- Band-reject filters remove frequencies a certain distance around a given frequency.
- Band-pass filters remove all frequencies except those a certain distance around a given frequency.

Any other filter types are variations on these themes. Subtractive synthesis is generally the process of using filters to carve out the type of sound the designer wants from a larger spectrum. The patch we'll talk about next uses a slightly simpler and more direct approach to building a sound. *Additive synthesis* is the process of adding to an empty palette the frequencies the designer wants. Each of these types of synthesis has its place, and they're not mutually exclusive.

A Toast!

For the next example we'll spend some time analyzing a recorded sound and then building a reproduction of it. In the sample-code download directory, in the sound subdirectory there is a file called Knife-on-wineglass.wav. This is a recording of me striking the side of a crystal wineglass with a knife. Play it now a few times and try to describe what you hear.

This wineglass strike is a very distinctive sound. It could be used as the basis for a notification alert sound; as a sound effect in a restaurant scene; or wherever a piercing, striking, bell-like tone would work. The main goal here is to analyze the original sound and see what parts of it we can reproduce. A secondary goal is to parameterize the patch so that changing a single value, the fundamental frequency, will change the sound but keep the characteristics of the sound in proportion to that value.

Analysis

Let's start by taking a look at what's going on in the recording. We'll look at a *spectrogram* of the sound, which shows which frequencies are most active. Various sound applications will help you analyze a sound in this way. I used Adobe Audition here.

We want to look at the frequencies' strength to see how we might be able to reproduce enough of the same frequencies to make a sound similar to the tap on the wineglass. Figure 9, *Spectral Display of Wineglass Strike*, on page 58 is a screenshot of Audition's Spectral Frequency Display view of the recording. Have a look and pick out generally where the strongest frequencies occur.

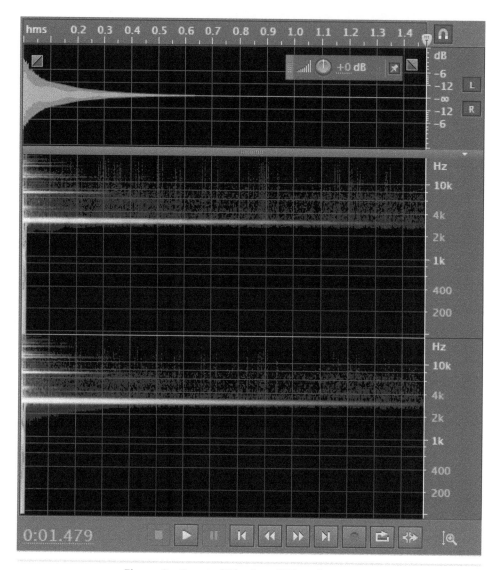

Figure 9—Spectral Display of Wineglass Strike

The figure shows three panes vertically. The top pane, the waveform view, shows the sound's amplitude over time, and is the most common visualization of a sound. It shows that the sound starts out with a lot of energy and fades out over the period of 1.5 seconds.

The following two panes are more interesting to us now because they tell us the spectral content of the sound, which means they show us which frequencies are most active over time. There are two panes because this sound was recorded with a stereo microphone; the top is the left channel and the bottom is the right. The graph's x-axis is time, the y-axis is frequency, and the intensity of color in the graph shows the amplitude at the corresponding frequency on the y-axis at that time.

Looking at either pane shows the same thing: there's a handful of frequencies that are very active and other frequencies where nothing at all is happening. One frequency, around 3 KHz, is by far the strongest.

The frequencies fade out at different points in the recording. The one at 3 KHz is pretty much silent by about 1.4 seconds, the second strongest (a little above 6 KHz) is gone a little after 0.3 seconds, and the ones higher than that are gone even more quickly.

The last thing to note is that there is a quick burst of energy across the whole frequency spectrum at the beginning for 2 three 3 milliseconds. That's the impact of the knife on the glass. The rest of the sound is the vibration of the glass after the strike. If you look closely, though, you'll notice that there are two of these full-spectrum events. My hypothesis is that the knife rebounds off of the glass and strikes it again in the space of 1 to 2 milliseconds here.

Take note that this method of analysis and reproduction doesn't work for all kinds of sounds. This kind of bell-like tone clearly has a set of active frequencies and is easy to reproduce by this method of synthesis. Still, this is a good opportunity to dig into a sound and see what's going on, and then reproduce it by copying its behavior.

Approach

The approach to building a patch to reproduce this sound is to break down the sound into a set of frequencies that behave like we see in the spectral display in the preceding figure. Since the frequencies decay at different times, we'll use different envelopes to fade out the frequencies at different times. The example patch uses only three frequencies, but you could keep adding more and more, with diminishing returns, to make the patch sound more and more like the recording.

For that burst of energy at the beginning of the sound, it's easiest to use a quick burst of white noise. It covers the full spectrum and does a great job of sounding like a quick impact or click. As an exercise and to add a little more realism, we'll reproduce that quick little rebound with a delay~ to retrigger the noise.

To complete the second goal of making this patch more reusable, we'll have a concept of a fundamental frequency that the other frequencies are derived from. This sound has a tone to it, unlike waves or wind, which have active frequencies across the entire spectrum. In a sound like this one, the *fundamental* frequency is the dominant frequency. Frequencies above the fundamental frequency are called *overtones*. The set of frequencies including the fundamental and the overtones are called *partials*.

If a sound's overtones are whole-number multiples of the fundamental, they are called *harmonic* overtones; for instance, 880 Hz is the second harmonic of 440 Hz. Overtones can be fractionally related to the fundamental, too. The more complex a sound is, the more complex the fractional relationship of the overtones to the fundamental.

To pick the partials to reproduce, we'll use the clearly dominant frequency from the spectral diagram as the fundamental, then find the next two strongest frequencies and figure out the relationship to the fundamental. Using Audition and a little careful listening, I've broken these down as follows:

- A fundamental of 2911 Hz
- The first strongest overtone of ~6841 Hz, or 2.35 * 2911
- The second strongest overtone of ~11061 Hz, or 3.8 * 2911

Approximating the durations of the frequencies, we'll use these fade-out times:

- Fundamental: 1 second
- First overtone: 300 milliseconds
- Second overtone: 200 milliseconds

And finally, we'll have the volume of each in proportion to the click at the beginning:

- Fundamental: 75%
- First overtone: 70%
- Second overtone: 60%

Now let's look at the implementation.

The Patch

For this patch, we'll create four groups of objects in similar configurations: one group for the attack's noise and one for each of the three partials. An ADSR envelope will control each group. The overtone groups will be almost identical to the fundamental group, but they will multiply the fundamental frequency by the ratios listed previously. Then a message box with two messages will trigger the envelopes and set the fundamental frequency.

Let's start by creating the attack group.

The Attack

Create a new patch in the directory you're using for this chapter, and start out by making the group that generates the attack portion of the sound, like the following image.

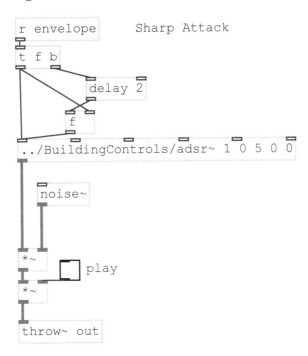

This group receives the envelope message and passes it to a trigger. The trigger passes the message on as a float to the ADSR, which jumps to 1 in 0 milliseconds and decays to 0 in 5 milliseconds—very quickly. This lets out a short burst from the noise~.

The trigger also sets off a delay, which waits 2 milliseconds and then triggers the envelope again, to simulate the two quick bursts of energy we saw in the analysis of the sound. This is overkill, perhaps, but it's fun to model little subtleties like that.

The Partials

Now let's create the fundamental group, pictured in Figure 10, *The Fundamental Group*, on page 62. Remember that the fundamental is the first of the partials. Add the objects pictured here to the patch.

The group labeled Fundamental passes the fundamental frequency from the message fundamental directly into the osc~ and is multiplied by the ADSR. To be able to play around a little and test out how the different partials sound,

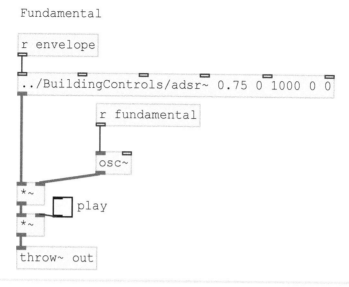

Figure 10—The Fundamental Group

there's also a check box controlling another *~. Turning the check box on and off will turn the group on and off.

Now add the overtone groups by making two copies of the fundamental group, one for each partial. For each, add a *~ between the r fundamental and the osc~, with the argument set to 2.35 for the second partial and 3.8 for the third. The ADSR arguments should be

- 0.7 0 300 0 0 for the second partial
- 0.6 0 200 0 0 for the third partial

That makes the second 70% as loud as the fundamental and fading over 300 milliseconds, and the third 60% as loud and fading over 200 milliseconds.

Finally, each group feeds its output into throw~ objects with the name out. Those throw~ objects send the signal over a named channel, which is received at a catch~ object using the same name. That's a nice pattern to use to cut down on having connections all over the screen.

The Completed Patch

The following figure shows the completed patch. Notice the catch~ out object, which receives the signals from each of the throw~ objects. It then feeds it into a *~ 0.25 to turn the volume down one-quarter. We need this because each group is sending out signals at full volume, so we need to turn the total amplitude down by the number of groups.

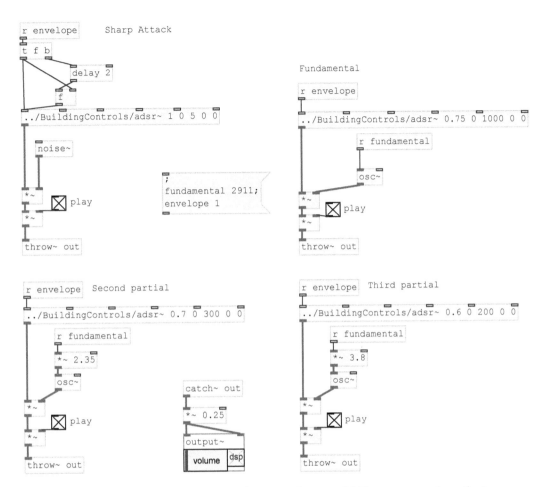

To complete the patch and get everything working, add the message box that sends the envelope 1 and fundamental 2911 messages. With relatively little work in Pd we've reproduced a real-world sound with pretty striking accuracy, if you'll pardon the pun.

Things to Try

As always, playing around with the controls is a great way to learn. Try a couple of things:

Turn the different "play" check boxes on and off and see what each partial does to the overall sound.

Try changing the fundamental and see how it affects the sound.

Next Up

I hope this was a satisfying chapter for you. After learning a lot about Pd and sound in general in previous chapters, it's good to be able to actually make some sounds that are remarkably similar to real-world ones. It's surprisingly easy using the tools we have in Pd.

As we created our effects in this chapter, we saw some new objects in Pd: the noise~ white-noise generator and the lop~ and bp~ filters. We saw how to load subpatches from other directories, and how to organize patches by using subwindows. We discussed how the wind and waves patches use subtractive synthesis, while the wineglass patch uses additive synthesis. Finally, the wineglass patch was an introduction to a more technical type of sound analysis and reproduction.

Next chapter we will cover additional interesting techniques and skills in Pd before we put those to use creating even more effects.

CHAPTER 5

Working with Waves

In the last chapter we had fun putting what we've covered to work with some practical effect examples. Now let's look at a few more techniques and skills that will be useful for creating some more interesting effects in the next chapter.

When you are done with this chapter, you will

- Know how to export and import sound files
- Understand the fundamentals of how digital sound is captured and stored
- Hear geometric waves other than the sine wave
- Know some other important synthesis types

Let's get started by learning the useful skill of being able to export a sound file from what we're hearing in Pd.

Exporting Sound Files

Pd has a lot of ways to tell the computer how to make sound, but what about recording the sound? Pd does have a way to capture the audio flowing out of it. It's up to us to tell it when to start and stop recording. Let's do that now. We'll take the wineglass patch from the previous chapter and extend it to add a feature where it records each time we send the message to make the *clink* sound.

Although Pd is primarily a *dynamic* sound environment, there are occasions where it's useful to capture a static sound from Pd. One case may be to simply preview some sounds at your leisure outside of Pd. Another may be to capture a few performances where variables are changing randomly, and then review them later. Another, which we'll encounter in a later chapter, is to produce sound effects for a static environment where we can't use Pd, like in a browser. We'll use what we cover in this section to export sound effects built for our web project.

Adding a Recorder to a Previous Patch

To get started adding a recorder, make a copy of the wineglass patch from the last chapter. Add a message sending a bang to record at the end of the main message: record bang. There's nothing special about the message name record; it just describes well what we want it to do for us. Then add a receive object for that message, and below that create a subwindow with an object called pd recorder. The main patch will look like the following figure.

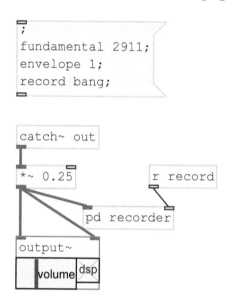

The subwindow contents look like Figure 11, *Recorder Details*, on page 67. There are two inlets, one a signal and one a bang, and everything is built around the object at the bottom, a writesf~ object. The argument to the writesf~ object is 1, which tells it to write one channel, making it a mono file, which works fine for this patch.

A writesf~ object will write a signal into a sound file and is controlled by a set of messages. The signal from the subwindow's leftmost inlet is fed into the left inlet of writesf~. The control inlet on the right is fed to a trigger that controls three messages. The first message (remember that a trigger triggers from right to left) is an open message with one flag and an argument. The argument is the relative path to the filename to write to, and here we have it hardcoded to clink.wav. That means when the sound data is written to disk, a new file called clink.wav will show up in the same directory as the patch. The -bytes 2

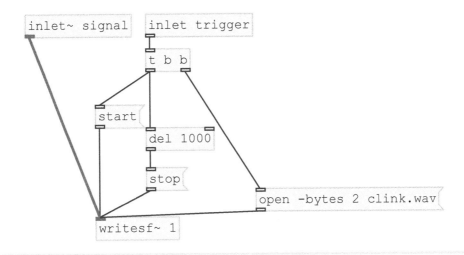

Figure 11—Recorder Details

flag and argument tells the open message to specify opening the file for writing as a 16-bit file. We'll talk more about that in a second. It's possible to determine the file type exported by changing it to one of the following: .aif, .wav, or .snd.

The open message to writesf~ is immediately followed by a start message, which tells the object to start writing its signal to the file. The bang message to start also kicks off a delay, which waits a second and then calls stop. This is a hardcoded way to write a second of signal data to a sound file.

Once you have everything wired up this way, make sure the receive bang is connected to the subwindow's inlet, and give it a try. Press the message and then listen to the WAV file that should appear in the same directory as the patch.

writesf~ Options

Let's talk a little about the options available to you when writing a signal into a sound file. These options allow you to make tradeoffs between quality and file size. Deciding what you need to export can help keep file size down, which is important when dealing with a medium like the Web.

Choosing Number of Channels

First of all we have the argument to the object writesf~, which specifies the number of channels to write. In the case of this patch, one channel is all we need, because there is no need for stereo data or anything more complex than that. In later patches we'll use panning, but here we can make the sound file we export smaller by specifying only one channel to writesf~. More channels make the file bigger.

Choosing Bit Depth

The next option is the one to the open message. The -bytes flag takes an argument of 2, 3, or 4, which controls the file's *bit depth*. These arguments correspond to a bit depth of 16, 24, and 32 bits, respectively. The higher the bit depth, the more information that can be encoded per sample. This sounds good, but how much do we actually need? Let's take a brief look at what goes on when we record sound.

Recording Digital Sound

Two factors determine the quality, or rather fidelity, of recorded digital sound: bit depth and *sample rate*. I distinguish between quality and fidelity because it's not always the case that more is better, so don't think that lowering these settings will necessarily produce a lower-quality recording. Choosing appropriate recording settings can keep *enough* of the signal and help keep file size down.

Understanding Bit Depth

A recording's bit depth controls the *dynamic range* it can capture, while the sample rate controls the frequency range. The dynamic range is the distance between the softest and the loudest sound that can be captured per sample. In most cases the default setting to the -bytes flag, 2, which corresponds to 16 bits per sample, is perfectly adequate. CD-quality audio has a bit depth of 16. If you happen to be making music or effects with very loud and very soft sounds, perhaps you could explore exporting at 24 or 32 bits, but for most cases, sticking to 16 will be fine and will keep file sizes down.

Understanding Sample Rate

To complete the story of how digital recording works, let's look at sample rate. Digital sound recordings, unlike analog recordings using audio tape, are not continuous representations of the signals they capture. In other words, digital recordings must capture a signal in discrete chunks, or samples, so that they can be stored in the digital medium. The rate at which these chunks are captured, or sampled, determines the fidelity with which a recording can capture different frequencies. If the rate is too low, a high enough frequency can take place in the gap between samples and be mistaken for a lower frequency, which doesn't sound good. This is called *aliasing*. Therefore, sample rate is most important for capturing high-frequency sounds with good fidelity.

How high a rate is high enough? According to the Nyquist-Shannon sampling theorem,[1] it needs to be at least twice the rate of the highest desired frequency. Since, as we saw earlier, at most the highest frequency humans can hear is 20 KHz, a sample rate of 44,100 Hz is commonly used as a baseline for a high-fidelity sample rate. CD audio is 44.1 KHz. Sometimes digital recording equipment goes even higher, which can help make sure the fidelity is maintained when moving between different equipment or software. In Pd the sample rate can be changed in the global Audio Settings menu.

Audio export is a useful capability when working with Pd. Understanding a little about how digital audio recordings work will give you the power to maintain the balance between a faithful recording and a lower file size. Now that we've seen how to export audio files, let's look at how to pull audio recorded in files into Pd.

Loading Sound Files

Now that we've seen how to export sound from Pd, let's look at loading sound from external files. Again, although Pd is primarily a sound-generation tool, there are plenty of occasions where it's useful to load external sound from files as a starting point to synthesis. Sometimes it's better to use a recorded sound that captures a sound with high fidelity than to try to re-create a sound with some other type of synthesis. In fact, a few forms of synthesis are built on loading and processing sounds from memory or from external files. These external files or sounds are called *samples*. It's perhaps unfortunate that the word *sample* is overloaded and ambiguously refers to both recorded sound files and a specific value captured during digital sound processing, but context is usually enough to allow you to distinguish between the two usages.

Synthesis Methods Using Samples

One of the synthesis methods that use samples as a starting point is called *wavetable synthesis*. Wavetable synthesis uses prerecorded single-cycle samples of periodic waves instead of a signal-generating oscillator to make sound. It was created as a way to make interesting musical sounds that were complex but still efficient enough to work on synthesis hardware with low processing power. Wavetable synthesis generally uses processes from other types of synthesis, such as additive and subtractive synthesis, to sculpt the sound as needed.

---

1. http://en.wikipedia.org/wiki/Nyquist%E2%80%93Shannon_sampling_theorem

Another method of using samples is called *sample-based synthesis*. The difference between wavetable and sample-based synthesis is strictly that wavetable synthesis uses a single cycle of a sample as a period of a periodic wave, where sample-based synthesis uses the a longer sample as a whole—for instance, to reproduce the sound of a piano from recorded samples of different piano notes. Sample-based synthesis also uses other synthesis methods to shape the sound as needed.

Other synthesis methods, such as *granular synthesis*, use samples, but really the difference between all these methods is how much of the sample is used and what the goal of using the prerecorded sound is. The common element is the prerecorded sound. Let's look at how to load samples into Pd a few different ways. In the next chapter we'll use what we cover here to load a sample of an everyday sound and modify it to something more epic.

Playing a Sound File with readsf~

The first method of reading a sound file is readsf~, the partner to writesf~, which we used previously. The readsf~ object uses the same messages to control it, but reads a sound file instead of writing to it. To try it out, create a patch and add the objects and messages from the following image.

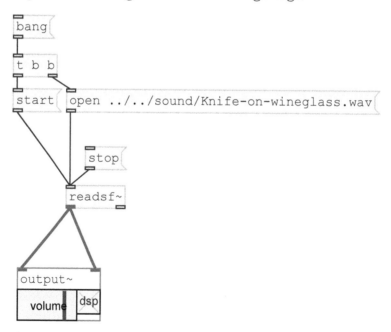

The patch starts off with a bang, which triggers two messages, an open and a start. Remember that trigger objects go from right to left, so the open message is triggered first, and the start next. The open message takes a sound file as an argument. In this situation the patch is relative to the sound file according to the path in the message, so make sure your patch is configured with the correct path to a sound file. The patch pictured here loads the wineglass-tap sound and plays it each time the bang is pressed.

The open message knows how to open a few different types of sound files by understanding the extension. There are some more advanced arguments in case you need more control over opening the sound file. The important thing to note here is that the open message needs to get to the readsf~ object before the start message. After the sound file is done being played, readsf~ holds no record of the file from the message, so the open needs to be issued each time. We can also use a stop message to stop the playback at any time.

Although it's not noticeable for small files, opening the file does take time, so you might need to preload large files. Finally, readsf~ takes an argument for a number of channels, and an argument for how big of a buffer to use. The object will have an outlet for each channel, starting on the left, with one more outlet, the rightmost, used to send a bang when the sound file is done playing.

Using readsf~ is a good approach when playing samples straight from a file, when you don't need to change much about the sound. It also could be useful when switching between a set of samples. Another characteristic of this approach is that it uses less memory since the file's contents don't stay in RAM.

Loading a Sound File with soundfiler

A different approach with more flexibility at the cost of keeping a sample in memory is to load the sound file's contents into an array. Previously we used an array to graph an oscillator; this is another useful way to use one.

Create a patch in the same directory as the previous example using readsf~ so that it can load the same sound file relative to it. Instead of reading a sound file and playing it immediately, this patch loads a sound file and puts its contents into an array. In the patch, create an array and name it wineglass. Then re-create the rest of the objects and messages from Figure 12, *Patch for Loading a Sound File*, on page 72.

The soundfiler object does the work of loading the file, but it is controlled by the read message sent to it. The arguments to read are first an (optional) -resize flag, then the sound file's location, ending with the name of the array in which to load the contents. The -resize flag tells soundfiler to resize the array to match the size of the file. This is useful if you don't know the size beforehand.

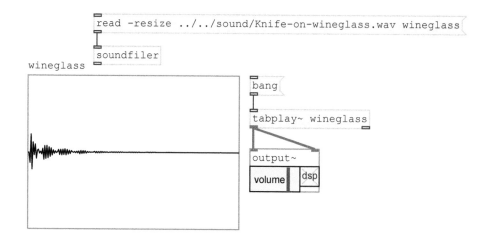

Figure 12—Patch for Loading a Sound File

Once the array is loaded with the contents of the sound file, we can use a different object to play the array's contents: tabplay~. The tabplay~ object simply plays the contents of the array named in its arguments and sends the signal to its outlet. The rightmost outlet sends a bang when its playback is complete.

In contrast to the readsf~ approach, loading a sample into an array uses a constant amount of memory as long as its loaded, but once it's loaded the sample can be played back as soon as it's needed. It's also possible to modify the contents of the array, process and send to a new array, or use the tabplay~ object's more advanced features to play parts of the array. Also, the soundfiler object can write data into sound files. Look at its help for more information.

Finally, we can choose which file we want to load into an array using an openpanel object. When openpanel receives a bang to its inlet it will open a system-file open dialog, and when the user chooses a file it will send the file path to its outlet. Figure 13, *A Modified Patch with $1*, on page 73 shows a modified patch using a dollar-sign variable in place of the hardcoded file path, which will allow you to choose which file to load with soundfiler.

We'll use the soundfiler approach in the next chapter to load in a sound file and play it back in a few different ways as an example of sample-based synthesis. Now let's have a look at creating samples of our own by generating data into an array.

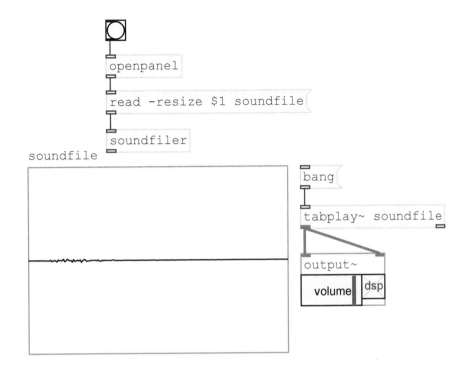

Figure 13—A Modified Patch with $1

Generating Wavetables

In the last section we covered how to load a sampled sound into an array for playback. This technique is the basis of sample-based synthesis. In this section we'll look at a different way of loading an array with data to play back: wavetable synthesis. Again, wavetable synthesis is where a single period of a wave is stored in a table, and then played back to create a wave with certain characteristics.

The benefit to wavetable synthesis is that it is a resource-cheap method of producing waves with interesting spectral characteristics without generating them in real time. So far in this book, when generating waves we've only varied the frequency and amplitude. Our perception of these two attributes of sound is part of how we distinguish one sound from another. Spectral content is another important way we can tell sounds apart. In a musical context the characteristics of a sound determined by spectral content are called its *timbre*, which is pronounced *tamber*.

As we'll discuss in the next chapter, being able to choose from waves with different spectral content, or timbres, allows us to craft sounds to our needs. A great way to make these waves in Pd is to generate them into an array at load time and play them back as needed.

Fourier Series: Creating Sounds from Simple Waves

When we synthesized the clink of the wineglass, we saw that adding together simple sine waves, or partials, allowed us to make a more complex, real-world sound, an approach called additive synthesis. In theory, by using additive synthesis we could synthesize any possible sound this way. Due to work of a mathematician named Jean-Baptiste Fourier,[2] we know that periodic signals can be broken down and described by sines and cosines, called Fourier series.[3]

The practical application is that by adding sine waves together we can generate waves other than the simple sine waves we've been generating so far with the osc~ object. If we do this when the patch is loaded, we can fill an array with this data and then play it back using a simple oscillator built for that purpose, and we can generate waves with widely varying timbres. These waves can allow us to build many different types of sounds.

Making a Sine Wave by Other Means

To see how to use this technique, let's build a sine-wave wavetable and play it back at 440 Hz instead of playing a sine wave using an osc~. Create a patch called sine_wave_wavetable.pd. Then choose Put > Array, name the array "sine," and choose not to save the contents of the array. Then create a loadbang connected to a message with the two messages sine sinesum 1024 1 and sine normalize 1, starting with a semicolon (;) to signify custom messages, as we've done before. You should have a patch that looks like the one in Figure 14, *Building a Sine-Wave Wavetable*, on page 75.

The messages use two functions of Pd that act on arrays. Let's look at each of them.

The sinesum Message

The first message uses a function called sinesum, which takes two arguments: a number of points to generate into the array, and a list of space-separated numbers that represent the amplitudes of partials to generate. In other words, the first in the list represents the *fundamental*, the second represents the *first overtone*, and so on. The message starts with the table name, so in total

---

2. http://en.wikipedia.org/wiki/Jean-Baptiste_Joseph_Fourier
3. http://en.wikipedia.org/wiki/Fourier_series

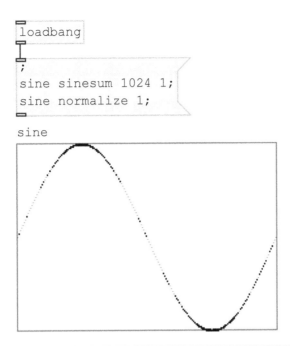

Figure 14—Building a Sine-Wave Wavetable

the message sine sinesum 1024 1 means "Generate into the array named sine 1,024 points worth of data describing 1 sine wave with an amplitude of 1 as the fundamental."

The Normalize Message

The second message tells the array named sine to *normalize* itself to 1. Normalization is the process of adjusting the array's contents so that the greatest value becomes 1 and the rest are adjusted proportionally. This makes sure that after we generate the wave the highest value doesn't exceed 1 but also is no lower than 1, so the sound is as loud as it can be without exceeding the limits of the system when playing it back.

Let's look at the sinesum function again. Its purpose is to generate a Fourier series of sine waves added together. Since in this case we only passed a list with one number, 1, to the function, it generated only one sine wave. Now we have an array containing one period of a sine wave. All that remains is to have some way to play the data out of that array as a wave, and we can have an alternative to the osc~ object.

Wavetable Playback

To play back the contents of an array, we can use an object called tabplay~, which basically sends the data in an array straight through to its outlet once. This technique works fine for sample-based synthesis, where we expect to play the data in the array as a sound. In wavetable synthesis we expect the data in the array to be one cycle of a periodic wave, so to play it back as a wave we need something that will continuously play the array's contents at the speed we choose. The object we want here is called tabosc4~.

Add a tabosc4~ sine connected to an output~ to the previous patch. This new object will act as an oscillator that takes data from an array. The frequency it plays at is sent to its inlet, so, as the following image shows, the patch also needs a number connected to the tabosc4~. Notice in the following figure how the message box also has a new first line with a 440, so it now does double duty by sending the frequency we want to the tabosc4~ through the number box as well as sending the initialization messages to the array. Turn the volume up on the output~, and you should hear the 440 Hz sine wave you've come to know and love.

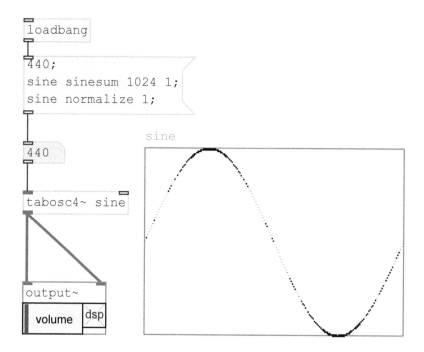

Now before we generate more interesting waves, let's address a few more things about this new patch. The first is why there is a 4 at the end of tabosc4~, and the second is why we used 1024 as the first argument to sinesum.

The tabosc4~ object has a 4 at the end of it to signify that it uses 4-point polynomial interpolation as it converts a constant number of data points from an array into a signal at different frequencies. Basically this means that it will do its best to smooth out the signal as it jumps from point to point of data in the array, no matter at which frequency the oscillator is generating a signal.

This is directly related to why we passed 1024 to the sinesum function. The help for tabosc4~ says that it expects the size of the array that it plays to be a power of two with an additional three points, one at the beginning and two at the end. The additional point at the beginning should be the same value as the final point in the array, excluding the two additional end points, and the final two should be copies of the first point in the array, excluding the additional beginning point. This is so that the tabosc4~ can make a smooth transition between each period of the wave. Luckily using sinesum or cosinesum takes care of this for us, as well as resizing the array.

When we specified 1024 as the size of array to generate, sinesum actually generated the extra transition points and added them to the array, resizing it to 1027. We used 1024 because it gives us plenty of room for enough points to make a nice-sounding sine wave. The help entry for tabosc4~ suggests using an array of size 512 for any generated wave with up to 15 partials. When generating above 15 partials it suggests using 32 times the number of partials and then rounding that up to a power of 2. Since we generated only one partial, 1024 is twice as big as the Pd manual says it needs to be. This could be a tradeoff between superhigh quality and memory.

Now let's look at generating some different waves with many more than just one partial.

Synthesizing Other Geometric Waves

Although lots of different waves could be generated using the sinesum technique or other techniques, let's generate some other *geometric waves* beside the sine wave. Geometric waves are basically waves with a geometric shape. We'll generate a sawtooth wave, a square wave, and a triangle wave, which any musicians interested in synthesizers will recognize easily. The following sections will show you how to reproduce these waves.

Sawtooth Wave

The sawtooth wave, sometimes called a ramp wave, is an interesting contrast to the sine wave. Whereas the sine wave is a pure tone consisting of only the fundamental frequency, the sawtooth wave is a very harmonically rich wave.

A sawtooth wave's partials are in the ratio of the inverse of the harmonic number. In other words, the fundamental partial is present in a ratio of 1:1, the second partial is present at 1:2, the third at 1:3, and so on. This is what makes the wave so harmonically rich: all the harmonics of the fundamental are present, just attenuated down as they get higher.

The sound of a sawtooth wave is sometimes described as "buzzy" and "bright," especially compared to the pure tone of a sine wave. It's a great wave to use with subtractive synthesis because it contains so much harmonic information even after using filters to carve out the frequencies we don't want.

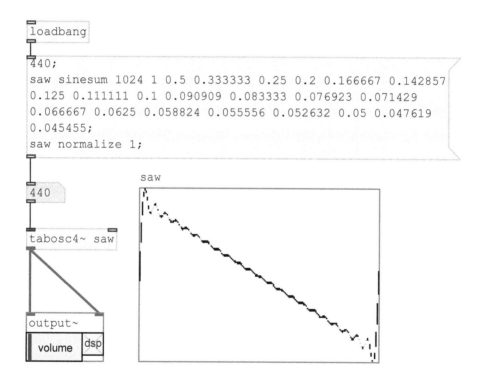

This is an image of a patch with a generated sawtooth (or saw) wave. Notice that the graph shows a wave that is a little more ragged-looking than an ideal straight line from top left to bottom right. This is because we're approximating the wave here by adding a number of partials together. The harmonic content is still easily recognized as that of a sawtooth wave. After creating a patch for the sawtooth wave, turn up the volume on the output~ and notice how much brighter it sounds than a sine wave.

With the sine wave we used only one partial to the sinesum function, but here we used 22. We specified 1024 points, which is still plenty for that many partials. Instead of calculating that many partials and entering them, you can use a bash shell script provided in the code download for this book (in utility/sinesum). The syntax (assuming you execute it from the root of the code download) is utility/sinesum [saw, square, or triangle] [number of partials]. An example interaction is outlined here.

⇒ **utility/sinesum saw 5**
❮ Amplitudes: 1.000000 0.500000 0.333333 0.250000 0.200000

This should make it a little easier to generate partials!

Square Wave

Now let's generate a square wave. Whereas a sawtooth wave contains every partial, square waves contain only the odd harmonics above the fundamental, while still in the same ratio to the harmonic number. For example, a square wave has the fundamental at a ratio of 1:1, skips the second partial, has the third partial at 1:3, skips the fourth, and so on. Here's a sample sinesum utility interaction.

⇒ **utility/sinesum square 5**
❮ Amplitudes: 1.000000 0 0.333333 0 0.200000

Square waves don't have all the harmonic content of a sawtooth wave, but are still pretty rich. In contrast to sawtooth waves, they can be described as having a hollow sound. Figure 15, *Patch for Square Wave with 22 Partials*, on page 80 shows a patch with a generated square wave with 22 partials.

Play the square wave and contrast what you hear to the sawtooth wave and the sine wave.

Triangle Wave

The last geometric wave we'll generate is the triangle wave. It's slightly more complex than the previous two waves. It contains only the odd harmonics above the fundamental, but they are more attenuated than with a square wave—the harmonics are proportional to the inverse of the *square* of the harmonic number. Also, every other odd harmonic is negative. For example, the fundamental is present at 1:1, the second partial is not present, the third is –1:9, the fourth is not present, the fifth is 1:25, the seventh is –1:49, and

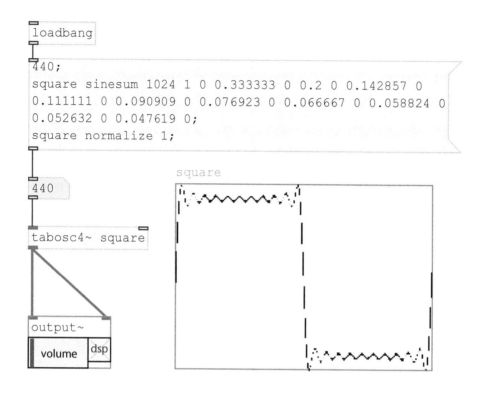

Figure 15—Patch for Square Wave with 22 Partials

so on. Figure 16, *A Generated Triangle-Wave Patch*, on page 81 shows a generated triangle-wave patch.

The sound of a triangle wave is like a slightly more muted square wave—still with a fairly rich harmonic content, although a bit more dull, and with a more hollow sound than a sawtooth wave.

Next Up

In this chapter we covered some practical skills, like how to export a sound file from what we generate in Pd, how to load a sound file and play it back, and how to generate data into an array for later playback. On the more theoretical side we covered how digital sampling works, including concepts like *bit depth* and *sample rate*. Then we talked about the difference between sample-based and wavetable synthesis. Finally we wrapped up with the fresh new sounds of a few geometric waves.

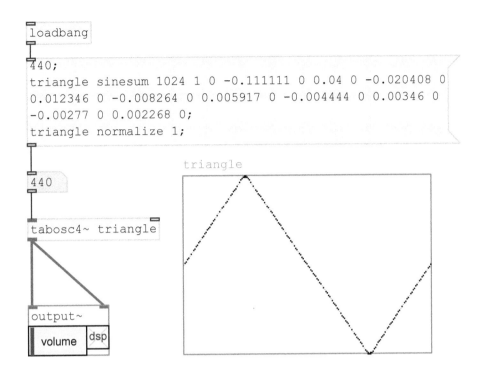

Figure 16—A Generated Triangle-Wave Patch

Things to Try

As always, playing around with the controls is a great way to learn. Try a few things:

Can you figure out what partial series would generate a wave something like the wineglass example? Of course, in that example the partials reached zero amplitude at different times, but see what you can generate with sinesum.

Try a few examples using cosinesum instead of sinesum. What do you notice is different about these generated waves?

When we built the low-frequency oscillator (LFO) a few chapters ago we used a sine wave from an osc~. What might be different about the LFO if we used a different geometric wave like one we generated here?

Next we will put what we covered in this chapter to work with real-world examples, one using sample-based synthesis and the other using wavetables.

CHAPTER 6

Creating Effects: Swords!

This is the second and final chapter where we create effects based on what you've learned so far. We'll have some fun with these! We'll create two effects, one modeling something that exists in our world, and the other modeling something that exists in a sci-fi world.

The first patch will be the sound of two swords clanging together in a sword fight, using a sample. The second will be the sound of a lightsaber from the *Star Wars* movies. That one will use a wavetable approach.

When you are done with this chapter, you will

- Understand how to dynamically manipulate a prerecorded sample loaded into Pd

- Use one of the waves we discussed in the last chapter to create a realistic sci-fi sound

- Know how to add smoother, more realistic dynamic effects and panning to your patches

These are going to be some fun patches! Let's dig in and see how to build them.

A Sample-Based Effect: Swords

Let's say we're designing the sound for a new native mobile game for iOS and Android, *Swordhaver II: The Search for More Things to Hit with a Sword*. It's a game with sword fights, and we want the sounds the weapons make to be realistic. We want to have sounds for lots of different situations where swords are clanging together, or off of armor or shields, not just one or two samples we play over and over. Having some variation will help the game's sense of realism and immersion, but we don't want to pack the game full of sound files and have to ship a huge game and manage a bunch of assets. Although it's not necessarily the only way to proceed, we'll use a sample-based approach.

For this patch, we'll focus on one type of effect, the sound of two swords clanging together. We'll build a patch that will allow us to have a few variations in sound, duration, and stereo panning, so the sword strike sounds like it's coming from a different part of the field of view in front of the player (assuming the game is in first-person perspective).

Analysis and Approach

The best thing to do if we want to capture a realistic sample of swords striking each other is to record two swords striking together. I've always wanted a sword, but my wife has never let me have one, let alone two. But really, what's a sword but a relatively thin bar of metal? An alternative could be to record a sample of two metallic things striking together and manipulate it as needed to produce a swordlike sound.

In the code download, alongside the wineglass sample, in the file named sound/single-knife-hit.wav is a sample I recorded of two kitchen knives striking together. We can load this sample into a table when Pd loads the patch, and then play it back at a lower pitch to make the knives sound more like swords striking.

To do this, we'll just play back the sample at a slower speed. This technique is the simplest approach we can take to change the pitch of the sample from a high-pitched knife sound to a lower-pitched sword sound. If we play the sample at half the speed, every frequency in the sample sounds twice as low. This works for our purposes because small physical bodies, such as knives, vibrate faster when struck than larger ones, such as swords. There are more complex ways of changing a sample's pitch to make the pitch change while keeping the length the same or to keep pitch the same while changing the time, but in this case I think it sounds better for our purposes to use the simpler technique.

We're also going to have a way to control the sound's duration so that we can have shorter and longer strikes. This is a way to add a little variation, and could be tied to a game variable. We also want to have a way for the strike to sound like it happened on the right or left side of the stereo field—the panning, which could also be tied to a game variable.

The Patch

The patch will load the knife-striking sample and present a few test messages to test out different pitches and durations. Instead of having a variable to directly control the panning, for this test patch we'll randomly pan the sound somewhere in the field to get the feel for what it might be like in the game.

The Main Patch

Create a new patch and save it somewhere handy. Copy the sound/single-knife-hit.wav sample into the same directory, or wherever you'd like it to be. Next create an array named "clang," a loadbang, a soundfiler, and a message to read the sample into the array—for example, read -resize ../../sounds/single-knife-hit.wav clang. Adjust the message for the relative path to your sample. You will have a patch that looks something like the following image.

Below the soundfiler is a send object, s samples $1. When soundfiler is finished loading a file, it sends to its bottom outlet the number of samples read from the file. We capture this in the send object's dollar-sign variable, and send it as a message called samples. The number of samples read into the array is important information we'll use to play back the array's contents.

Next we'll create a set of test message boxes for different values that we'll use to control the playback. The variables we'll use for the control parameters will be as follows:

- *pitch* will control how fast we play the sample, with a value of 1 being normal speed.
- *envelope* will trigger an envelope to play the sample.
- *duration* will control how long we want the sample to play.

In the patch from the sample code, as the following figure shows, there are four message boxes for a few different examples.

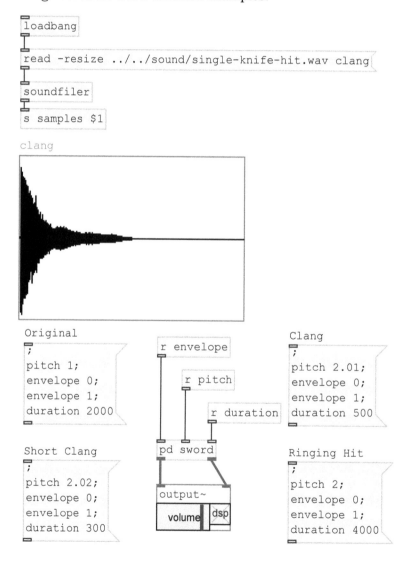

One plays the sample as is, as a reference, and then three more play it at around twice as slow, with slightly different durations and variations in speed.

Create those messages and then the r objects for each control variable as in the preceding image. The patch so far works as sort of a control layer, and pushes the mechanics of playing the sample into a subwindow abstraction. Create that subwindow now with an object containing pd sword. In the messages before we send envelope 1 we first send envelope 0 to stop any previous sound.

Controlling the Sample Playback

Inside this subwindow we'll create the objects to play the sample back at the speed that we want, as well as control the duration of the playback. First let's revisit what it takes to play a digital sample.

We've loaded the contents of a .wav into a Pd array. That file was recorded at a sample rate of 44100. Remember that the sample rate is an indication of how many samples were taken every second. The number of samples in the file was calculated by the soundfiler object and broadcast as a message called samples, allowing us to capture that number for use in our calculations.

If we want to play back the samples from the file at the same speed it was recorded, we need to send 44,100 samples to Pd's output every second. To do that we need to know the number of samples in the array and how long the original sound was. Since we know the number of samples and the fact that the sound file was recorded at a sample rate of 44100, we can divide the samples by 44,100 to determine the length in seconds. Then, if we want to change the speed at which we play the sample back, we can vary the length by a factor to adjust the playback speed. The contents of the subwindow in Figure 17, *Varying the Length of the Playback*, on page 88 do just that.

First of all, there are three inlets corresponding to the three control parameters: envelope, pitch, and duration, with the first two connected to an 'adsr~' envelope. Those inlets will receive input when the message boxes are pressed. The r samples objects will receive input as soon as the sample file is loaded, though, and the left receive object will store the number of samples into the adsr~ inlet that controls the attack initial value. The right receive object divides the number of samples by 44.1 and stores it in a *~ object, where it will scale the incoming pitch value.

It's not strictly necessary that this patch use an adsr~ envelope; we could have used a line~ object and a set of other objects to control it, but the adsr~ takes up a little less space, and it is really a set of line~ objects inside the subpatch anyway.

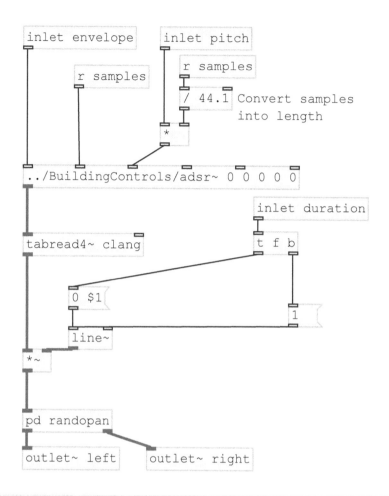

Figure 17—Varying the Length of the Playback

The goal of the top half of this subwindow is to control the tabread4~ in the middle. A tabread4~ reads from an array, which contains a table of data. It uses four-point interpolation, which basically means that it knows how to wrap around seamlessly to the beginning when it reaches the end of the table. Here it will read from the "clang" table, but it needs to be told how fast to read. We do that by using the envelope to tell it to read from the array sample by sample. The envelope will be initialized with its attack value set to the number of samples in the file by the r samples receiver.

Then, when it receives a 1 message sent into the inlet envelope, it will ramp down to 0 over a certain number of milliseconds. That number of milliseconds will be based on the number coming into the inlet pitch. We want the pitch control variable to be 1 for a normal playback speed. The number calculated and stored in the * object will be the percent of the speed at which we want the file to be played, so if the pitch message is something other than 1 it will scale the playback time accordingly, making the adsr~ ramp smoothly from the number of samples down to 0 in that amount of time. As the envelope ramps, it tells tabread4 which sample to read from the array.

The inlet duration will get the duration we want the sound to play and use a trigger to send a message to a line~. First it initializes the line~ to 1 by sending 1 to its right inlet, and then it sends a message to ramp down to 0 in the number of seconds specified to the duration inlet. This doesn't do anything fancy to the sample playback by tabread4~; it simply fades it out because it's being multiplied by the *~.

Finally, the signal passes through the randopan abstraction, which we'll talk about in a second, and gets sent to two separate left and right outlets. Create all the objects described previously using the images as reference, and then create a subwindow with pd randopan, and we'll see how we have Pd randomly pan the sword-strike sound between the stereo channels.

Random Pan Location

Panning is the process of placing a sound in a *stereo field*. As you probably know, stereo sound is produced from a right and left speaker, and helps us perceive sound as naturally coming from a certain direction. Pd produces stereo and provides us with a right and left inlet on objects such as dac~ and output~. When a sound is more present, or louder, in the right channel, we perceive it as coming from the right. The final subwindow in the patch is the random panning feature. This will randomly pan each sword strike it receives between the right and left channels to simulate giving some spatial depth to the sound the user could correlate to visuals onscreen.

Inside the subwindow create an inlet signal object and two outlet objects, called left and right. The subwindow splits a signal into two, with a randomly calculated amplitude for each, such that when added together they equal 1. Create the rest of the objects you see in Figure 18, *Varying the Length of the Playback*, on page 90, and we'll discuss what roles they play.

Each time the envelope message is received, we calculate a new random number between 0 and 100 with the random 100 object. Then we divide that by 100 and send the value to the left outlet. Before we send the value to the right outlet we calculate the inverse with the swap 1 and the subtraction object (-). This is the technique that we saw before in the low-frequency oscillator (LFO)

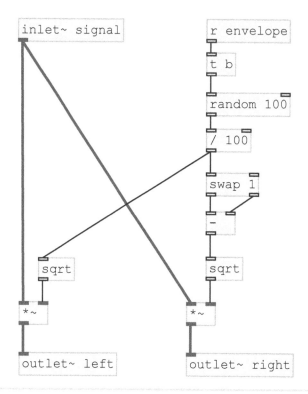

Figure 18—Varying the Length of the Playback

subpatch. The signal is multiplied by these numbers and sent to the outlets. Now the left and right outlets will have some portion of the signal passing to them, randomly balancing the full signal between the two.

Before mixing in the sides together with each other using a *~, we take the square root of each side with sqrt~. This is a better approach than simply splitting the signal in two, because if we did that, then the sound would actually be *half* as loud when panned to the center. This may be counterintuitive, but it has to do with the fact that the actual loudness is the square of the amplitude. Mixing in the square root of the other side's signal leaves more amplitude when panned near the center, and addresses this issue better. This is one strategy for building a panner. For a more in-depth look at the theory and a few different strategies for building controls like panners, have a look at Chapter 14 in *Designing Sound [Far12]*.

Wrap-Up

This patch nicely creates the sound of two swords clanging together with a fairly simple technique, and gives you an example of how to get much more control over a static sample.

Now let's look at a more complex, even more interesting patch that models an iconic sound that should be instantly recognizable: a lightsaber.

A Wavetable Effect: A Lightsaber

A long time ago in a galaxy far, far away (Hollywood), a sound designer named Ben Burtt created the iconic sound of the lightsaber for the first *Star Wars* film.[1] That humming, buzzing sound is instantly recognizable, and the electric crackle of two lightsabers in close contact lent an extreme, visceral tension to those saber battle scenes.

Lightsabers don't exist, but onscreen they sounded *exactly as you'd think they should*, which made them completely believable. That's excellent sound design. No one could accuse us of setting our sights low by taking on the sound of the lightsaber; it's our last stand-alone sound effect before we turn to creating effects for specific platforms.

There's a lot going on with a lightsaber that we could try to model, but for this patch, let's stick with three elements:

- The activation, or unsheathing/sheathing
- The idling hum once it's activated
- The higher-pitched, louder hum of the swing

That's plenty for a good-sized patch. We'll get to see in action a few techniques we've talked about in previous chapters, and you'll learn about a few more that will come in handy as you make your own sound effects.

Analysis

How Ben Burtt designed and discovered the lightsaber sounds is a relatively well-documented piece of sound-design lore. Searching around the Web, you'll find the details in a number of places. Here's a great video of Ben himself describing the lightsaber-design process.[2]

The sound is a combination of two sources: a recording of a warm, buzzing movie projector and a recording of the interference caused by a television

---

1.   http://en.wikipedia.org/wiki/Ben_Burtt
2.   http://www.youtube.com/watch?feature=player_embedded&v=i0WJ-8B6aUM

induced into a microphone. The swinging hum of the lightsaber cutting through the air was the result of playing that same sound back over speakers while recording it using a mic swung back and forth in front of the speaker in time with what was happening onscreen. This produced a realistic, dynamic sound, and because of the Doppler effect the frequency increases a half a tone or so.[3]

For my preparation I downloaded a bunch of lightsaber sound effects from the Web, listened to videos, and read an online tutorial on re-creating the lightsaber sound without the aid of a digital signal processing tool like Pure Data.[4] I used the frequencies discussed there, verifying them using frequency analysis in Adobe Audition and by using my own ear. The patch will have a lower-frequency buzz at 38 Hz, and a higher-frequency electric hum at 92 Hz. When it re-creates the swing it will pitch up the hum to 97 Hz.

Approach

To describe the approach the patch takes, I'll explain each part we're planning on reproducing. Don't worry if each bit isn't clear in isolation; if the description is unclear when you read it, listen to the part it describes and hear what it's describing.

To re-create the higher, humming sound, we'll use a square wave. A sine wave won't work because it's too basic and boring. A ramp wave is no good; it's too rich in harmonics. A triangle wave is rich in harmonics, too, and dull-sounding. A square wave, with its odd harmonics, sounds right for an electric sort of hum. To re-create the low buzzing sound, we'll use a custom wave. For both of these, we'll use a wavetable.

To give the idle lightsaber the effect of a little motion we'll use a *chorusing* effect, which I'll explain in a moment. For the unsheathing and sheathing sound, we'll use white noise and an effect called *flanging*.

For the swing we'll use the square wave sound of the idle at a higher frequency, louder than the idle sound, and distorted a bit. Let's look at the patch and explore how each part is modeled.

The Patch

Let's dig right into Pd. You can certainly try to follow along and re-create this patch from scratch, but remember that the completed patch is in the code

---

3.   http://en.wikipedia.org/wiki/Doppler_effect
4.   http://www.dblondin.com/071807.html

downloads. I recommend playing with the completed patch first to hear each part so the sound is in your head as we dig into how it's made.

Here is an image of the completed patch.

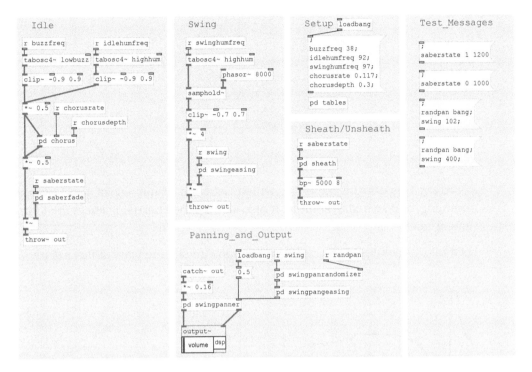

Since this patch is pretty complex, I've used canvases to visually separate the different sections. These are the gray boxes with titles.

A Word About Canvases

To create a canvas, use the Put menu or the key sequence ^⇧C. Then right-click in the top-left corner and select Properties. It's not simple to resize the canvas the way you want, but playing around with the width and height values and pressing Apply will eventually get you the right size. To get a canvas to sit behind objects already on the screen, you may have to cut and paste the objects since the only way to control the foreground order is to change the creation order.

Now let's discuss what's happening in each of the sections separated by the canvases.

Setup and Testing

In the section titled Setup, pictured in the following image, we set some initial values and build the wave tables used in the rest of the patch.

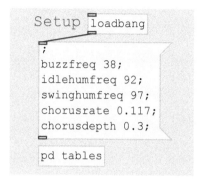

A loadbang triggers the message, which sets the frequency of the buzzing, humming, and swinging sounds. It also sets up the chorusing effect the idling section uses.

Let's take a look inside the table subwindow to see the wavetables set up.

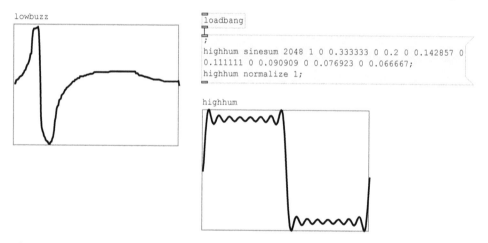

The highhum table is set up with a square wave using the technique from the last chapter. The lowbuzz table is set up using a new technique, though.

As you may remember from when we first talked about arrays, there's an array properties option labeled Save Contents. This will cause any data in the table to be saved along with the patch. Here I turned that option on and drew a wave shape into the graph using my mouse. You can use this technique to create custom waves.

At a low frequency, this wave gives a nice mechanical, buzzing sound. Its shape is interesting because it has a large positive and then negative jump, but then settles down for the remainder of the cycle nearer to the middle, or zero. Because of this burst of energy and then the rest of the period with little energy, it causes a discontinuous tapping sound at very low frequencies, like 1–5 Hz. As it gets higher in frequency the buzzy character stays with it, so it works perfectly for our goals. If you'd like, try re-creating a wave like this and using a tabosc4~ to experiment.

Finally, next to the Setup section, there's a section labeled Test_Messages (Pd doesn't like spaces in its canvas titles). These messages (pictured here) show how we'll control the patch's functioning.

We'll activate and deactivate the lightsaber by sending the saberstate message a list of values. The first element of the list will be a 1 or a 0 for turning the lightsaber on or off, and then a value indicating how long we want this to take, in milliseconds.

When we want the lightsaber to swing, first we want to generate a random panning value, similar to what we did with the sword in the patch in the previous section. Sending randpan bang will do this. Then we'll send a value in milliseconds to swing to say how long we want the swing to take.

Now that we've seen the patch's controls, let's look at the first sound you'll hear as you activate the lightsaber. I've called this part of the process sheathing and unsheathing.

Activation: Sheathing and Unsheathing

The goal of the sheathing and unsheathing section is to replicate the whooshing *shhhhhhhhhhhht* sound of the lightsaber activating and deactivating. As the sound progresses it gives the listener a sense of motion taking place, and when it completes it stops. We accomplish this by sweeping a signal up or down against some white noise, creating *interference*. Let's set the stage and then dig into what that means. The following is the sheath section of the parent patch.

First we have a receive object, which gets the list that the test message saberstate sent. That list is passed into a subwindow called sheath, which outputs a signal into a band-pass filter and into a throw~. The band-pass filter is at 5 KHz, with a fairly high Q value of 8. Remember, the higher the Q value the narrower the passed bands the filter will let through.

I added this band-pass filter to focus the sound of the unsheathing and sheathing into the higher frequencies, and arrived at the values by adjusting to taste.

The Sheath Subwindow

Let's look into the subwindow and see how the sound works (Figure 19, *Lightsaber Sheath Subwindow*, on page 97).

First, in the top right, there's an inlet for the saberstate message. The incoming list is passed to two different subwindows, saberline and sabertoggle. The first abstraction's job is to create a signal to sweep the sound up or down, creating the impression of sheathing or unsheathing. The second simply turns the sound on when it's sweeping up or down, and then turns it back off when the sweep is complete. We'll look into these abstractions in a second; first let's see how this sweeping signal works.

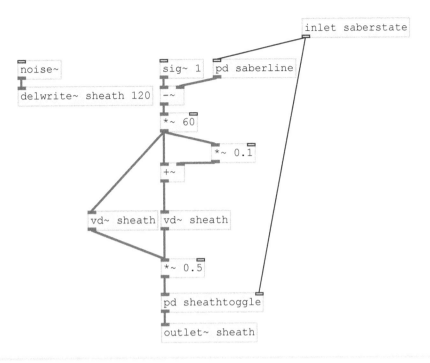

Figure 19—Lightsaber Sheath Subwindow

The main component of the sound is the noise~ on the left of the patch. The signal is fed directly into a delwrite~. This object is one component in a *delay line*. A delwrite~ reads a signal into a named memory location, in this case named sheath, and specifies a value in milliseconds. The named memory can then be read by either a delread~ or vd~ object, at a later specified time, up to the amount of time in milliseconds specified by the delwrite~—in this case 120.

This allows us to delay a signal and play it back after it originally was played. With this arrangement you could create an echo effect by playing the original signal, and then a short time later play whatever was played once again.

The Whooshing Sound: Flanging
In this case we don't want an echo. Instead we want to sweep through the noise's signal and combine it with another slightly delayed copy. This creates interference. Here's how interference works: when two opposite signals are combined with each other they cancel each other out. In other words, if the amplitude of the first signal is 1 while the second's is –1, the result of combining the signals will be 0, or silence. This is the principle behind noise-canceling headphones.

If the signals are both amplitudes of 1 at the same time, they would double the amplitude. However, if the signals are *close* to being the same amplitude at the same time because one is slightly delayed, then a very interesting effect happens, still due to interference. Some bands of the audible spectrum cancel each other out, which filters out a set of frequencies. If the length of the copied signal's delay is changed over time, different bands of frequencies are canceled out, which creates a whooshing sound. The effect of this process is called flanging.

For a visual indication of what's happening, look at the spectral analysis of a recording of this patch.

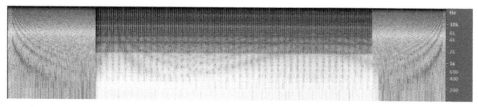

In this window, just as we saw in the wineglass analysis, the graph shows frequency distribution over time. The lightsaber is activated, then left to idle for about 5 seconds, and then deactivated. Notice how there are sets of darker descending curves on the left and ascending curves on the right, which look a bit like ripples on a sand dune. Those are the unsheathing and sheathing sequences, and the ripples are the frequencies that are being canceled out due to interference. Look at this image while you use the test messages to turn the lightsaber on and off, and you can visualize the effect.

Achieving the Sweep

Now that we have a better understanding of the theory behind what we're trying to achieve and how it works, let's see how we achieve it in the patch. Referring back to the image of the sheath subwindow, you'll notice that the output of the saberline subwindow is a signal that's fed into a signal-subtraction object, -~, into which is also fed the output of a sig~ 1, a constant signal of 1. Let's look inside that subwindow now (Figure 20, *Lightsaber Sheath Saberline Subwindow*, on page 99).

This abstraction creates a smooth transition from the leftmost value of the incoming saberstate list to its opposite value, so if the saberstate message comes in as 1 1200, saberline produces a smooth transition between 1 and 0, taking 1200 milliseconds. As the list comes in the inlet at the top, we use an unpack with two float arguments, which splits the list into two values and sends each to two separate outlets. Then we use a select object connected to the unpack's left outlet, which will be 1 or 0. The select object with arguments of 1 and 0,

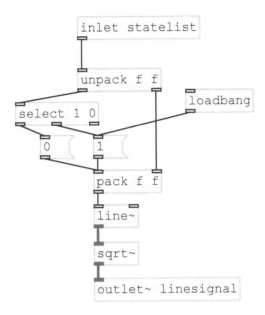

Figure 20—Lightsaber Sheath Saberline Subwindow

in that order, sends a bang to the left outlet if it receives a 1 to its inlet, and a bang to the second outlet if it receives a 0. If it gets something else, it passes that value on to the third, rightmost, outlet.

Next, we assemble a list using a pack object. If the list 1 1200 came in, at this point we'll have a list containing 0 1200. That's sent to a line~. As you'll remember, a line~ interpolates a smooth transition between its current value and the first element of a list, doing so in the amount of time specified in the list's second element. Notice that we're assuming the list~ is always set with the correct previous value. Currently if we break this assumption and try to turn the lightsaber on when it's already on, telling the line~ to go from 1 to 1, we'll just hear a burst of noise for 1.2 seconds. We use a loadbang attached to a 1 message to set up the patch the way we want at the beginning.

After the line, we have a sqrt~ object. This eases the transition from the line~ to its final value a little more naturally than just a straight line. If you look at the lines described by the "sand dunes" in the unsheathing and sheathing sections of the preceding spectral-analysis image, you'll see the curve rather than the straight line. This is more natural-sounding because things in nature, including sounds, rarely describe straight lines. Our brains expect real-world sounds to ease into place. If you've ever dealt with visual animations, you've probably run into the concept of easing functions. The sqrt~ here acts as a simple, audible, easing function.

Now coming out of the saberline abstraction is a nice curve to sweep the whooshing sound up or down the frequency range. This curve drives how much the -~ at the top of the sheath abstraction subtracts from the sig~ 1. The value is then multiplied by a *~ 60, so we have a signal easing from 0 to 60 or 60 to 0, depending on which direction we're coming from.

Next in the signal chain are two vd~ objects. VD stands for *variable delay*, and as we saw earlier, these will read from the sheath delay memory at a certain point up to 120 milliseconds, controlled by the curve from the saberline abstraction. The leftmost vd~ is driven directly by the *~ 60, but the one on the right has a +~ and a *~ 0.1 before it, so the two variable delay objects are separated by an increasing distance as the saberline increases, and a decreasing distance as it decreases.

Toggling the Unsheathing/Sheathing Sound

Then the output of the two delays is combined and halved by a *~ 0.5 to bring the signal back to between –1 and 1, and then sent into the sheathtoggle abstraction. Let's look into what that does.

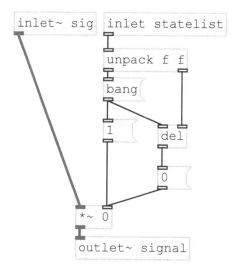

Again, the goal here is to turn the unsheathing/sheathing sound off when it's done. To do this it also takes an inlet from the saberstate message, but sent first through an unpack. No matter what the first value of the list is, it triggers a 1 sent to the left inlet of a *~ and kicking off a del that counts down to the millisecond value of the list, and triggers a 0. This multiplies the incoming signal by a 1 for the duration of the unsheathing or sheathing, and then multiplies it by 0, effectively turning it off when the sequence is done.

With this deceptively simple part of the patch explained, let's have a look at how to create the idling lightsaber sound.

Idling

The goal of the idling section of the patch is to re-create the humming and buzzing sound of a still but activated lightsaber. The section labeled Idle is pictured here.

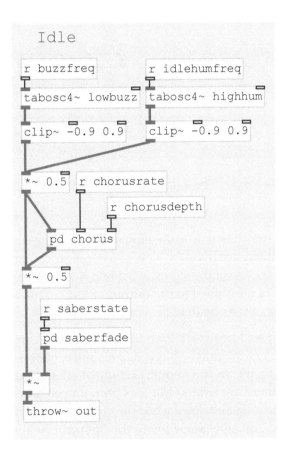

When the patch is loaded, messages are sent to buzzfreq and idlehumfreq, which set the frequency on two tabosc4~ objects, shown in the preceding image. These each point to the arrays, lowbuzz and highhum, that were initialized in the Setup section. These each correspond to the buzzing and humming sounds described at the beginning of this section; one each for the projector and the TV interference on the microphone.

Then, to give each tabosc~ interesting harmonic content, they're passed through a clip~ object, which limits the amplitude on either to between 0.9 and –0.9. This adds a little distortion to each sound, and is completely a matter of taste; I experimented with it and liked it. Then, after being combined and halved by a *~ 0.5 the signal is split, half going straight through into another *~ 0.5 and the other half being sent through another abstraction called chorus before rejoining the unaffected signal. Look at the chorus subwindow contents in the following image, and then I'll explain what's going on here.

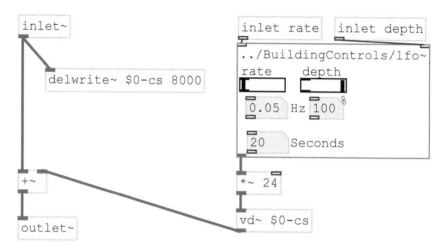

You can see that the chorus is quite similar to the flanging effect discussed earlier. The only difference between a chorus effect and a flanger is the length of the delay. Here the incoming signal is fed into a delwrite~, and then summed with a +~ instead of multiplied to the output of a vd~. This vd~ is swept back and forth with our LFO subpatch.

The effect is slower and less stark than the flanging effect on the unsheathing and sheathing part of the patch, but it has a similar effect—look back at the spectral-analysis image, in the middle section where the lightsaber is idling, and you'll notice there are similar but less pronounced "sand dune" interference patterns cycling up and down while the chorus is running. This produces a sense of motion, and adds some life to the lightsaber's static idling.

The LFO's rate and depth are set with the initialization messages to chorusrate and chorusdepth sent from the parent patch's Setup section.

This is a very simple chorusing effect, but it adds a lot to the sound. Let's look at the last abstraction the signal runs through before being sent to 'throw~ out', a subwindow called saberfade.

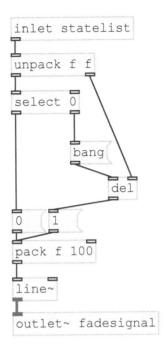

The goal of this abstraction is to fade the idle sound in after the unsheathe sound is complete, and fade it back out before the sheathing sound starts. It does this by unpacking the saberstate message to its inlet, using a select to decide if the first element of the list is a 0; if so, it uses a pack to send 0 100 to a line~, and otherwise it packs and sends 1 100.

This will fade the idle sound in or out over 100 milliseconds after the unsheathing or sheathing is complete. The control output of saberfade is multiplied by the idle signal and then sent to a throw~ out.

Next, let's look at how the patch works when we activate the swinging of the lightsaber.

The Swing

When the lightsaber swing was recorded for the movies, a louder recording of the idle sound was captured by swinging a microphone like a sword in front of a speaker. Because of the Doppler effect, this changes the tone a little, so we'll have the swing frequency higher, as we can see by looking at the setup message values.

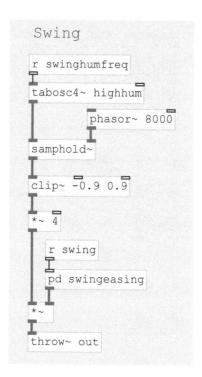

The initial frequency value is received from the swinghumfreq message into another tabosc4~ reading from the highhum array. I chose to use only the highhum array for the swing part of the sound, and not both, since this sounded a little cleaner to me.

The signal flows out of the tabosc4~ and into a samphold~ that has a phasor~ connected to its right inlet. The goal is to add a slightly harsher humming quality to the sound of the swing by adding "jagged" edges to the smooth changes of the square wave coming from the tabosc4~. The samphold~ accomplishes this by taking a sample of the signal coming into its left inlet and outputting only that sampled signal until a change is received on the right inlet.

The phasor~, which sounds slightly more awesome than it is, is a oscillator that describes a smooth, linear ramp from 1 to 0 over its period. It's very similar to the sawtooth wave we saw in the previous chapter, except that the sawtooth wave goes from 1 to –1. Phasors are often used for controlling other signals, and in this case it's set to 8 KHz, so every time it changes value from 1 to 0—8,000 times a second—it tells the samphold~ object to take a new sample. The easiest way to understand this may be to look at a graph of the signal at the different points in the signal chain.

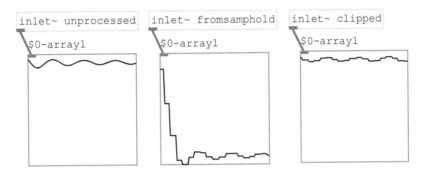

Here I hooked up our simple grapher~ abstraction to three graphs: the leftmost is coming out of the tabosc4~ directly, the second is from the samphold~, and the last is after the clip~. The graphs don't show a full period of the wave, but they show enough to pick out what the samphold~ does to the smoothness of the square wave. Now we have a slightly more prominent-sounding hum when the swing message is sent.

Back to the parent patch: after the clip~, the signal is bumped up four times louder with a *~. This is because we want a much louder swing sound. Keep in mind we'll have to account for this in the output section.

Finally, the signal is combined with the output of another abstraction, called swingeasing, shown in Figure 21, *swingeasing*, on page 106. This is to control the volume level of the swing in a more interesting way than ramping it up and down linearly.

We do this by taking the swing duration from an inlet and having that trigger a 1 to our ADSR (attack, decay, sustain, and release) envelope, which turns it on, but also setting the envelope's attack duration with the incoming value. Instead of the sqrt~ we've used to give a line a more natural-sounding curve, we use a slightly more complicated function, using Pd signal math objects.

$$\frac{x^a}{x^a + (1-x)^a}$$

Using 2 as a value for a means we can square the x and (1 - x) signals by multiplying them by themselves with a *~. When a is 2, this function describes a curve as in Figure 22, *Function Curve When a is 2*, on page 106.

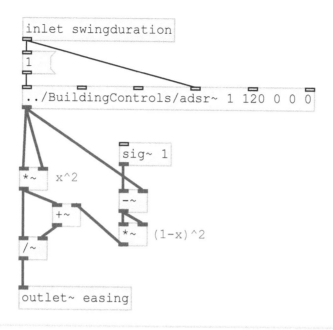

Figure 21—swingeasing

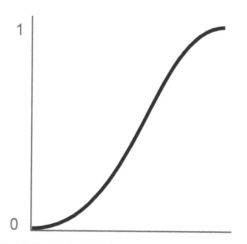

Figure 22—Function Curve When a is 2

This is a more natural easing into and then out of the swing. This signal is sent to the outlet and combined with the swing using a *~, and then the signal is sent to a throw~.

Output and Panning

Last in the signal chain is the section labeled Panning_and_Output.

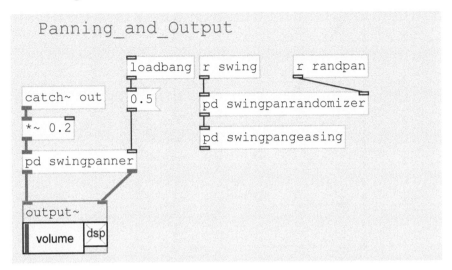

It starts with a catch~ out to combine all of the throw~ out signals, then sends that through a *~ to multiply the signal by 0.2, which makes sure that when combining the unsheathing/sheathing, the idle state, and the swing (which is multiplied by 4, remember) our highest amplitude doesn't exceed 1.

As shown in Figure 23, *The Pan Output Panner*, on page 108, before the final signal is sent to the outlet~, we send it through a panning abstraction similar to the one in the last section.

In this case we don't have the panning abstraction deal with the random pan location, though. For this patch we want the panning to change from its current location to a new random location, but only when the swing takes place, and over the duration of the swing. We also want the pan to follow the same easing curve as the swing. First let's look at the subwindow that deals with the random pan location, swingpanrandomizer, shown in Figure 24, *The Pan Output Randomizer*, on page 108.

This abstraction receives a bang to its right inlet, which triggers the generation of a random number. This number is stored in the float object. The left inlet receives the swing duration, which triggers a bang and a float, assembling the random value and the swing duration into a list using a pack f f object. This is sent out of the abstraction and into the easing abstraction, pictured in Figure 25, *Pan Output Easing*, on page 109.

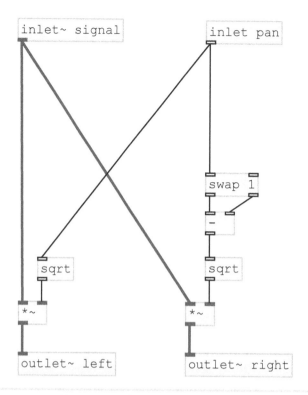

Figure 23—The Pan Output Panner

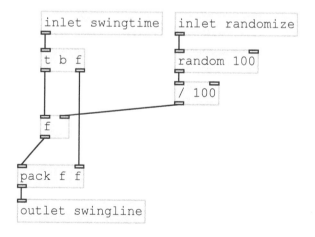

Figure 24—The Pan Output Randomizer

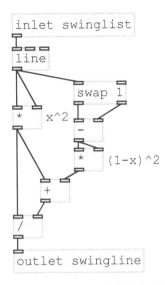

Figure 25—Pan Output Easing

The swingpaneasing subwindow simply contains a line to ramp using the list sent out of the previous abstraction. It then has equivalent non-signal objects that produce the same easing effect as the swingeasing abstraction from the Swing section. This curve is sent through the outlet into the swingpanner to ease the panning to its new pan location.

Wrap-Up

This patch was very detailed. It pulled together a number of techniques and tools we've covered so far, and introduced a few new ones. I hope it was fun to walk through, and to hear it, since the lightsaber is such a culturally iconic sound. This patch as it stands wouldn't hold up to being used in a movie soundtrack, but it does a good job of re-creating the sound recognizably.

The exercise of understanding how this sound could be modeled is interesting since a lightsaber doesn't exist in the real world, but we do have stories and notes on how the original sound was designed. One of the most fun parts of sound design is to model a sound for some sort of machine or situation that doesn't exist. You can achieve great results with creativity and an understanding of how things that *do* exist in our world work and how to use different wave forms to create certain types of sounds.

Next Up

In this chapter we've covered a lot of interesting ground even though we've discussed only two patches. We covered how it's possible to load in a sample and control the pitch by playing it back at different speeds. This is a great way to use Pd to gain a lot more control over static samples. Then we saw how to build a good set of the mechanics behind one of the most known and loved sci-fi sounds: the lightsaber. We used the wavetable techniques we covered in the last chapter, and picked out which wave would make the sound work the way we wanted.

In the next chapter, we'll take a step back from working with Pure Data and talk about the user experience of sound, which is as at least as important as getting the technical details right. This will be a short interlude before we wrap up with two projects using Pd to build sound effects for a web game and two native mobile apps.

CHAPTER 7

Sound in the User Experience

Now we'll take a quick break from building sounds to talk about an important part of sound design: the user experience (often called UX). Thus far we've been building patches with the sole intent of creating a realistic effect. Essentially we've been answering the question "Does this effect sound good?" without posing the question "Are we building an experience that sounds good?"

In the next two chapters we'll build projects with the goal of adding sounds to a web project and native mobile projects. In this chapter we'll take some time to talk about what makes the audible part of an experience worthwhile to the user. This is a valuable part of a sound designer's intuition; along with understanding how to make an effect sound correct, it's important to understand how to make an experience sound good as a whole, and to work well in the context of everything else the user might be hearing at the time.

When you're done with this chapter you will

- Have learned how to think about sound as part of an experience

- Know some important tips to keep in mind when delivering sound to the Web

- Know some important tips for delivering sound to native devices

Let's get started now talking about the user experience of sound in general.

Sound Is a Public Experience

Have you opened a web page or an app, and then had sound blast out of your speakers? Especially at the worst time, when you're in a quiet office environment and everyone's busy working, and in the worst way, when the sound was obviously and embarrassingly not work related?

This experience teaches a very important lesson about sound: by default, sound is a public experience. Although it's possible the user has headphones on, or maybe doesn't have a system that plays sound, it can never be assumed that sound won't be a public experience. In contrast, the visual part of a digital experience is generally private. Users are generally aware and protective of who can see what's on their screen, and it feels like eavesdropping to look at someone else's monitor even if it's in a public space.

Let's go through a few related general points that you should keep in mind when designing sound.

Your App Doesn't Need Sound

When designing your app (web or otherwise), try to start from the assumption that you don't need to have sound in it. In general this is the default position, but make it a conscious starting decision. It should be proven to your satisfaction that an app *needs* sound for a better experience; you should not include sound by rote.

One of the first things to consider when coming to the question of sound is the user's expectations. No user expects sound in a spreadsheet program. On the other hand, if your app has some sort of alarm or notification functionality, an alarm sound may be a user expectation. Of course, in general, the expectation for a game is that it has a sound.

By starting from a default position of having to prove the need for sound, and then considering user expectations, you'll have a better chance of making the audible part of your experience better.

Expect to Be Turned Off

If you do have sound in your app, you should never expect that users will passively let it play all the time. Users have a number of options available to control sound, sometimes at the speakers themselves, and definitely at the operating-system level. You shouldn't ever expect important information to get to users by sound alone.

Part of a user choosing to hear the sound in your app is a matter of trust. You'll have to gain that trust by letting the user control the sound; allowing sound to play in a disruptive or unexpected way will destroy the trust before it's even built.

Don't Make Anyone Listen to You

Related to the last point is that you should never make the user listen to the sound in your app. If you have the ability to play sound despite operating-system or hardware settings, such as with certain sound modes on mobile devices, don't use it without heed to the context the user is in. Listening to sound should be a user's choice.

If the user is already listening to music, she will not appreciate your app trying to play music on top of that, or even worse, turning the music off. If you have an app that has important sounds essential to the purpose—for instance, a training application with important dialogue—then it's appropriate to play that dialogue by default. But always let the user be in control, and make sure that the user understands the *value* of the audible part of the experience in your app.

Make Your Sound Mean Something!

Closely related to the last point is that your sound should have a value. I remember back in the early days of the Web, when developers first got their hands on Macromedia Flash. It was so easy to make your web page a "multi-media" experience that all of a sudden every time a user clicked on something he had to endure a low-quality, high-volume, clicking sound. It was never clear what purpose this served besides to prove that sound could indeed be played on the Web. As a user I didn't need any extra confirmation that I had clicked a button; I had the feedback of my mouse coupled with visual feedback.

As far as sound goes, button clicks have very low value. Consider the value of whatever sound you want to add to your experience. Perhaps in the context of a game button, click sounds are slightly more valuable because they fit the theme you're building on, and reinforcement of that theme helps immerse the user. I still find them irritating, but my point is not to proscribe this or that type of sound; it's to encourage you to consider the value and the meaning of the sound you put into your apps.

Now that we've considered the user experience of sound in a general way, let's talk about sound on the Web.

Sound on the Web

The web has been around a while now, as opposed to, say, mobile apps. That means user expectations, when it comes to things like sound, are pretty well defined. Nobody likes unexpected sound on the Web, and hearing it causes anger and frustration instead of delight and trust.

Here are a few general principles to keep in mind when working with sound on the Web.

Never Play Sound by Default

I recently saw a feature in a beta version of a popular browser that would place an icon on any background tab where its web page was playing sound. This feature makes it easier for the user to track down and close the offending site. Even web-browser developers are thinking about the problem of unexpected sound as something to fix, not promote.

We've all likely experienced frantically trying to figure out where sound is coming from on a web page to shut it off. This type of behavior seems to come largely from aggressive marketers and is associated with less reputable sites. As a basic design principle, when it comes to the Web, sound should be off by default.

Provide Control to the User

Rather than launching an aggressive, covert sound assault, other websites go out of their way to disable sound by default, giving the user the choice to opt in. Consider Vine,[1] Twitter's short-form video site. By default, when viewing a video on Vine the sound is off, even though it's a meaningful part of the experience. Users can toggle an icon to turn the sound on if they choose.

This progressive approach puts the power in the hands of the user and avoids destroying trust by blasting out sound right as the video plays. The principle of providing control to the user is an important one to remember on the Web especially, since browsing the Web is often a relatively passive experience.

Have Fallbacks

Another way to give the user more control over the audible part of an app's experience is to break up the types of sound into different levels. This may make the most sense in the context of a game, but the principle could work on certain nongame apps, as well. For instance, in a game you could allow the user to control whether the background music plays separately from other sound effects. This would allow the user to listen to whatever music she wanted, but still get important sound feedback from the game.

For the most part, these principles apply not only to the Web, but to mobile apps as well. Let's talk about a few other important topics related to sound in mobile apps.

---

1.   https://vine.co/

Sound on Mobile Devices

As mentioned, a lot of the design principles in the preceding section apply very directly to the Web, but may also apply to mobile apps. The main difference between the Web and mobile in this regard is the attention of the user. Almost all mobile operating systems have only one foreground application and limited screen space, so it's likely that you have more of the user's attention than on the Web. Again, always consider context.

There are a few things to consider that are particular to mobile apps, however.

Expect Phone Calls

There are many kinds of mobile devices, but a great majority of these are phones. That means there is a very important audible experience users expect from their device: a phone call.

It may be that the operating system you're developing for forcibly turns off the sound when receiving a phone call, but don't expect this to be the case always. As you should anyway, regardless of sound, test your app when receiving a call. Make sure no sound of yours ruins the phone call. Make sure you get back in the right state when the call is complete.

Consider Sounds from Other Apps

When considering how important sound is in your app, also consider how to play well with other apps that produce sound. You should not start with the assumption that your app's sound is more important than any other's to the user. Test and make sure that the sound coming from your app works well when sound is coming from other apps.

For example, in a game my team recently worked on, we looked at the current audio environment and determined whether the user was already playing audio—for instance, listening to music on the device. If so, we automatically configured our game to turn off the background music but keep the effects on. That way the user doesn't have to get frustrated and fight our app to do what he wants to do: listen to his own music. He could still go to the settings screen and turn on our background music, which would shut off his music-player app, but that's a choice we left up to the user.

Engineer for Small Speakers

An important part of the context of playing sound on a mobile app is the hardware the sound plays on. It may be high-quality headphones, cheap earbuds, or small mobile speakers. Test your sounds on all of these and listen to what you hear. Maybe your effects can be adjusted to sound better on a wider variety of speakers.

Listen on Lots of Devices and in Various Environments

Closely related to the last point is that you should consider the wide variety of contexts in which your sounds will be played. The user may be on headphones in the office, on a subway, in the library, or outside. Any of these situations will have a different effect on what your user can hear or how your sound will be apparent.

Both this point and the last are general sound-engineering tips. Music producers always listen and adjust the songs they're producing to sound best in the widest array of contexts they can. You can learn from this, and make sure your audible experience is the best it can be.

Next Up

In this chapter, we've gone over a number of points that you should consider as a sound designer. We've seen that sound should be considered a potentially public experience, and we've covered what that means to the sound we design. We've also considered some principles that apply to web and mobile experiences. These boil down to respecting users, thinking about their expectations, and not surprising them but rather building their trust.

Keep these principles in mind as you continue your journey as a sound designer. In the next chapters we'll get a chance to design sound for both the Web and mobile platforms.

CHAPTER 8

Exporting Sounds for a Web Game

So far we've been building patches with fairly abstract goals, but starting with this chapter we'll build some practical projects where we not only design sound effects, but design them with more real-world goals in mind. In this chapter we'll look at the design of a web game, design a patch in Pd to create effects for that game, and then cover how to integrate the sounds into the code. This game will serve as an example of a project where we can't use Pd in the final software, so we need to export static sound files.

There are many ways to build a game on the Web, and lots of frameworks out there to make it easier to do so. There are also ins and outs to working with sound on the Web, depending on browser support, and even a number of JavaScript libraries to smooth out some of those wrinkles. In this chapter we'll develop sound for a game, which means we'll build a Pd patch, of course, but we'll also touch briefly on some of the issues developers have in dealing with sound on the Web, and introduce a few techniques and frameworks to help.

When you are done with this chapter, you will

- Understand what is meant by "8-bit sound" and learn one way to approximate it

- Have a more practical understanding of how to export sound files from Pd

- Know some of the pitfalls and limitations you'll face when delivering sound to the Web

- Have some resources at your disposal to deal with these limitations

Now let's get started by looking at the design of the game we'll integrate with.

Designing the Game

The web game in this chapter is an example of a case where you can't use Pure Data directly, and need to export sounds instead. For that purpose, really any game will work, such as tic-tac-toe or a jewel-based puzzle game. But we want something a little more audibly interesting; something to give us a sound-design challenge. We'll build a patch for a sci-fi space game because that's way more fun than tic-tac-toe. Let's have a look at the game and an interesting design constraint it has that will make building our Pd patch that much more fun: we'll make this game sound like those '80s console games you may have grown up playing.

Space Rocks! A Side-Scrolling Space Shooter

The game we'll design the sounds for is a side-scrolling space shooter. That means the player's perspective is from the side of the action, and the game progressively moves forward from right to left. It's called a shooter because the goal is to destroy obstacles by shooting at them while trying to avoid running into things. It's a classic style that you're probably familiar with. The following is a screenshot of the game running.

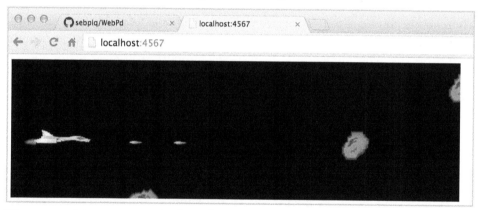

The spaceship can move in four directions to avoid the asteroids that fly in from the right of the screen. It can also shoot missiles to destroy the asteroids. In this simple example game there's no scoring or much in the way of a user interface to show things like lives remaining and the damage to the ship, but it can take three hits before it loses a life, and it has three lives.

It's a simple game, built as simply as possible, and really an object lesson in designing sound effects for a game. One other design goal affects how we'll design the sound for the game: it's designed to look and feel like an old 8-bit console game.

A Bit of Nostalgia: 8-Bit Graphics and Sound

Back when I was a kid in the early to mid '80s, the heyday of console games, a lot of these platforms shared a particular design characteristic: the 8-bit architecture. At the time having 8 bits to process graphics and sound information was state of the art, but nowadays we talk about 8 bits as a quaint limitation and make artwork and music with that limitation as a fun homage to the glory days of console games.

As we saw in the section *Understanding Bit Depth*, on page 68, when sound is stored digitally a few parameters affect its quality: the sample rate, which determines the maximum frequency that can be captured in a digital recording, and the bit depth, which determines how many levels of amplitude each sample can store. The fewer bits there are to store the amplitude of a given sample, the more quality will suffer. As an extreme example, if there were only 1 bit to store amplitude, each sample could only be on or off, and the sound wouldn't be recognizable.

Having only 8 bits was a limitation, but game sound designers were able to do amazing things within those constraints. Because of the system's limits, they couldn't have complete realism as a design goal. But by hinting at or approximating how things in the real world sound, or in other cases using sound effects that were more abstract entirely, they created really fun audible experiences.

This game is meant to look and sound a bit like one of those old 8-bit console games, so we'll shoot for that in the patch's design.

Introducing gameQuery

To build the game, I used an HTML game framework called gameQuery.[1] gameQuery is based on the almost ubiquitous jQuery JavaScript framework.[2] The game we use here is based on a gameQuery tutorial, and simplified and adapted to our needs. In some ways the game is a bit of an excuse to build the sounds, but with gameQuery it was easy to make a convincing web game with the feel of a side-scroller.

The Sound Effects

In the game, the player can move a spaceship in four directions and shoot missiles at asteroids. In the next section we'll build sounds effects for when the ship's thrusters fire when moving, the sound of the missiles firing, the

---

1. http://gamequeryjs.com
2. http://jquery.com/

sound of the asteroids being hit, and the sound of the ship being (unfortunately) hit by asteroids.

We'll build one patch to create all of these sounds, process them to simulate the sound of an old 8-bit game, and export the effects into sound files ready to be imported into the web game. Let's look at how we'll accomplish that now.

Designing and Building the Patch

The patch that produces the sounds for this web game is available in the code download in the pd/WebProject directory. If you'd like to build the patch alongside the description, you can follow along, but I'll stick to describing it here. In the following image, you can see the layout of the whole patch, grouped into sections with canvases like in the lightsaber chapter. We'll go through this patch section by section.

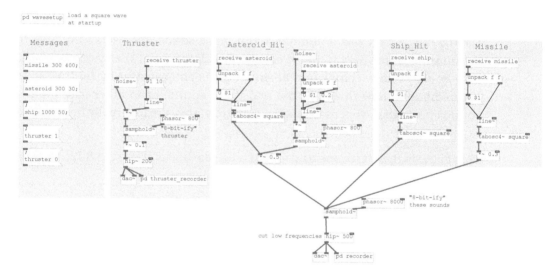

Similar to the lightsaber patch, there's a setup abstraction in a subwindow, and then a section for the control messages. For the sound effects, there's a section for the spaceship's thruster, one for the asteroid explosion, one for the ship being hit by an asteroid, and one for the sound of the missile firing.

Setting Up and Controlling the Patch

There's not a lot of setup for this patch. Most of the effects rely on an array that we set up in a subwindow, called square, and that is loaded at runtime with a square wave using the sinesum function that we've used a few times now. We're using a square wave because a lot of the 8-bit console systems of the '80s used sound generators that produced square waves, so to capture the distinctive sound of those systems the square wave is the best choice. The image here is from the pd square subwindow.

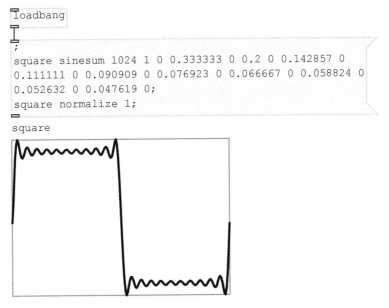

The following image shows the control messages. There's one message for each of the asteroid, ship collision, and missile sounds, and two for the thruster.

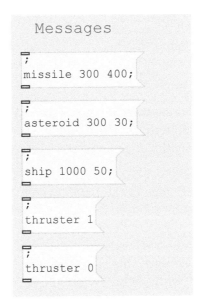

Besides the thruster message, which sends a 1 to turn the thruster on and a 0 to turn it off, each message sends a list with the first position corresponding to the duration of the sound and the second to a starting frequency. The basic idea of all of these effects is that a square wave tone will start at a certain frequency and fall to zero in a given duration. Let's look at each of these in more detail now.

The Thruster Sound

The thruster sound is controlled by sending a message to thruster: a 1 turns it on and a 0 turns it off. The following image is a close-up look at the patch.

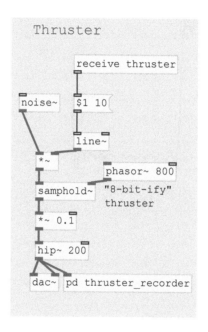

The sound generation is done with a noise~ object. The 1 or 0 value received from the thruster message controls a line~, which serves to smooth out the transition on or off by smoothly increasing the *~ object's value, by which the noise~ is multiplied.

After the noise is another stage of the patch, which creates the 8-bit sound we want for this game.

Simulating 8-Bit Sounds

In the lightsaber patch we used this exact method, where the sound is fed through a samphold~ object with a phasor~ controlling it. In that patch the phasor~ was set to a relatively fast 8 KHz, which added a little bit rougher quality to the sound. In this case, the phasor~ is set all the way down to 800 Hz, which will squeeze the random frequencies of white noise into a very squared-off random waveform. Again, this is only a simulation, but it's a convincing one. The following is an image with two of our grapher~ submodules.

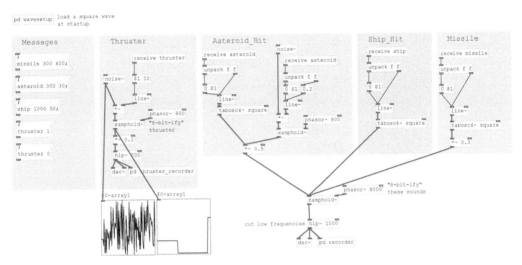

The graph on the left shows the output of noise~, and the one on the right shows the output of the samphold~ object. It's still random, but reduced to a much smaller set of amplitudes. It sounds very much like the rocket sound in a lot of old console games.

After the samphold~ there is a *~ 0.1 to turn the sound down a lot. Next we use a high-pass filter to cut out low-frequency sounds below 200 Hz and tune the sound a bit. Then the signal runs directly into a dac~.

Exporting a Wave File

Next to the dac~ there's an abstraction in a subwindow to help record the sound of the thruster into a sound file we can integrate into the web game. Figure 26, *Thruster Recorder*, on page 124 is a look at what's inside there.

This subwindow looks a lot like the recording abstraction we saw added to the wineglass patch in *Adding a Recorder to a Previous Patch*, on page 66. The signal inlet~ runs directly into a writesf~ that outputs a mono sound file. When the thruster message is received, we use a select object to decide if it's a 1 or 0. If it's a 1, a trigger object first triggers a message to set the name of the sound file to write to with an open thruster.wav message, and then sends a start message to writesf~. When a 0 is received, the select triggers a stop message to the writesf~ object, telling it to finish writing to the sound file.

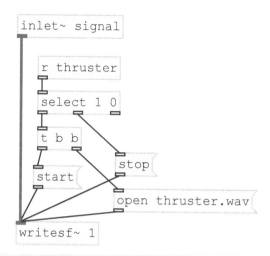

Figure 26—Thruster Recorder

If you've followed along building the patch, or just looked through the example patch in the code download, you can try triggering the thruster on and then off. When you're done you'll find a file called thruster.wav in the same directory as the patch with the sound of the thruster recorded in it.

The Hit Effects

Now let's look at the sections that produce the effects for the missiles, collisions, and explosions.

The Asteroid Explosion

When a missile strikes the asteroid we need a sound that works for the explosion of the missile and the disintegration of the asteroid. Figure 27, *Asteroid Hit*, on page 125 shows the sections labeled Asteroid_Hit.

There are two parts to this section. The right side is driven by noise~ and uses the same setup of a samphold~ controlled by a phasor~ set to 800 Hz as the thruster. This is to produce half of the explosion sound, the noisy, crackling, nontonal part. Its duration is controlled by receiving the asteroid message, unpacking that list, setting the starting value of the line~ to 0.2, a low value so as not to be too loud, and then picking out the duration element and sending the line~ down to 0 in that length of time.

The left side of the section uses a tabosc4~ driven by the wavetable with the square wave we saw previously. It's going to produce a sort of abstract *boom*

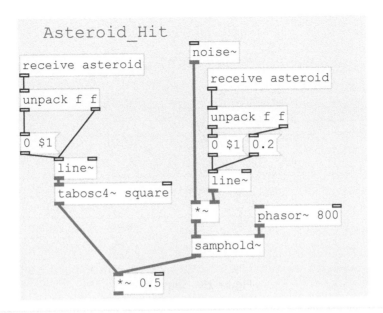

Figure 27—Asteroid Hit

sound by ramping from a low 30 Hz down to 0 Hz using a line~—the line~ controls the frequency instead of the amplitude of the tabosc4~ in this case.

First the asteroid message is unpacked, then the second component sets the value of the line to the starting frequency, and the first component is combined into another list, preceded by 0, to bring the line~ down to 0. 30 Hz is a pretty low frequency for people to hear, but because we're using a square wave and later processing the sound with another sample and hold, there will be harmonic overtones at higher frequencies to pick up on. Finally, the noise and the tonal component are summed and multiplied by 0.5 and the signal is sent on.

The Ship Is Hit!

The next section, labeled Ship_Hit, is identical to the left-hand-side, tonal component of the asteroid explosion, as you can see in Figure 28, *Ship Hit!*, on page 126.

The sound this section produces is the same, a falling tone over a given duration. The message for the ship collision uses a much longer duration of 1,000 milliseconds and a starting frequency of 50 Hz.

The sound design here could use a little discussion. What kind of sound best conveys the idea that the ship was hit, something the player doesn't want to

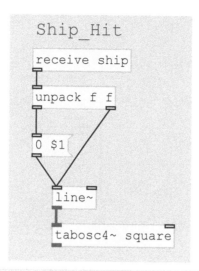

Figure 28—Ship Hit!

happen? I thought about maybe using a "red alert" alarm sound, but instead stuck with the same falling tone as the other effects, but with the duration longer for more emphasis.

The Missile Fire

The sound of the missile firing is produced by the section labeled Missile, shown in Figure 29, *Missile Fire*, on page 127.

This section is almost identical to the Ship_Hit section, except the signal passes through *~ 0.3 to get turned down a bit since it's a more prominent sound.

I chose to do this and other volume adjustment in the patch for simplicity's sake and because I noticed it needed to happen while building the patch, but it's worth considering whether to do this kind of adjustment in the patch or in the final game. It may be easier to keep this configuration out of the destination software, but it may also be better to have that kind of control in the game.

8-Bitifying and Exporting

The patch's final stage combines the output of the asteroid explosion, ship collision, and missile sounds into one processing and output area. The thruster is handled separately since its control messages work differently, allowing

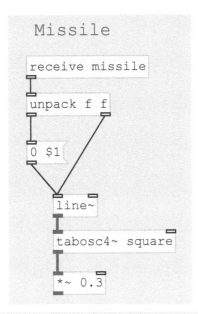

Figure 29—Missile Fire

you to choose how long the thruster effect sounds, but we can gain a little space and clarity in the patch by processing the rest together.

Processing the Hit Effect

As we can see in Figure 30, *Missile Fire*, on page 128, the first step to processing the sounds is to use the now-familiar samphold~ with a phasor~ stage to give the sounds the 8-bit quality we're looking for.

After that stage we put the final output through a high-pass filter set to 500 Hz, just as a matter of taste, to sound a little more like the 8-bit consoles of old. They were likely sending sound out of a small, mono TV speaker, which wasn't able to reproduce low frequencies very well, so a high-pass filter reproduces that effect.

Recording the Hit Effect

To record the various sounds to WAV files for use in the game, the output of these three sections and the processing goes into another abstraction, shown in Figure 31, *Web Group Recorder*, on page 129.

On the left side of the subwindow we have one signal inlet~ to take whatever sound is currently playing and direct it into a writesf~ for output to a sound file. The right side deals with using receive objects to decide which sound was

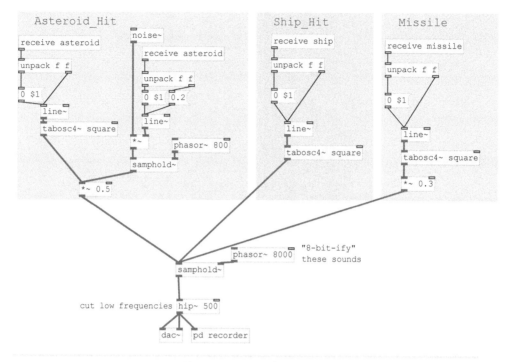

Figure 30—Missile Fire

triggered. If the r asteroid gets the message, it tells writesf~ to open a file named asteroid_hit.wav, and so on for the other messages.

Next, each receive is routed to an unpack f f so we can determine how long to make the sound file. We do this by using the first component of the incoming lists to trigger a start message to writesf~ and then start a delay set to that duration, which, when complete, triggers a stop message to tell writesf~ to finalize writing the exported file.

Now we'll move toward putting these exported sound files into the web game. First, though, we'll talk about working with audio on the Web.

Browsers: The Cause of and Solution to All of Our Problems

If you've done any web programming, you know there are differences in the various web browsers to overcome with code and design gymnastics. Dealing with sound in browsers is no different. This isn't a book about web programming, but since our project in this chapter delivers sound to the Web, it would be good to touch on some of the issues with sound on the Web before we go on to the code.

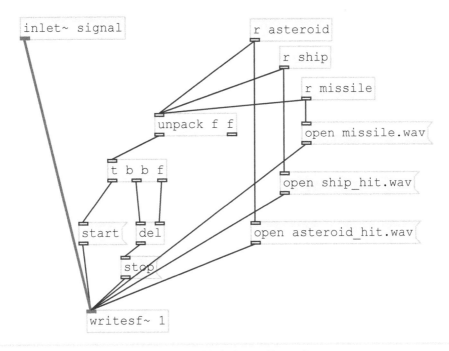

Figure 31—Web Group Recorder

Dealing with Browser-Compatibility Issues for Sound

It used to be that it wasn't worth dealing with sound in the browser, and most web programmers would opt to use Flash, which offered a lot more control over sound than any of the browsers. HTML has come a long way since then, but we're still not to the point where the same capabilities are guaranteed across all users' systems, even if they are fairly modern.

Web Audio: The Future

There's an interesting HTML spec called Web Audio.[3] It has a set of capabilities that go beyond just playing sound files, and actually allow developers to generate sound and control it with JavaScript. If you're interested in audio on the Web, you should definitely look further into Web Audio. In fact, a project called WebPd attempts to convert Pd patches to Web Audio.[4] Given those capabilities, it's a promising addition to HTML.

---

3.   http://www.w3.org/TR/webaudio
4.   https://github.com/sebpiq/WebPd

As you might expect, though, Web Audio isn't fully supported on all modern browsers. At the time of this writing, all WebKit-based browsers and Firefox support it, but Opera and Internet Explorer do not. For more information, keep up-to-date with a browser-capability catalog site such as Can I Use.[5] Until Web Audio is more universally supported, though, developers have to use static sound files.

File Types

One of the first choices to make when delivering sound to the Web with a sound file is which file type to use. Sadly, no format works for every browser on every system. One format to watch is the Opus codec,[6] which is an open audio codec that may gain support across browsers. At the time of writing, it's supported on only Firefox out of the box, and Chrome with a special flag. WAV files used to be universally supported, but they're large since they're uncompressed, and they don't work out of the box on Internet Explorer with the HTML5 audio tag. MP3 is a popular compressed format that's been around a long time, but for licensing reasons Firefox doesn't support MP3s out of the box on all operating systems. Ogg Vorbis is the open alternative to MP3, but doesn't work out of the box with Safari or Internet Explorer. Google's Chrome browser supports everything and works on all platforms, but the days of building websites that work on only one browser are, thankfully, long past. We haven't covered mobile browsers, which might even have sound disabled. What should a poor web developer do with sound?

Luckily for browsers that support HTML5,[7] the audio tag can contain multiple source tags, which allows the browser to pick the file format it can understand. That still means for full support on all browsers you'll need to provide versions of your sounds in WAV, MP3, and OGG formats.

Although HTML5 has developers covered by supporting multiple sources, not all browsers support HTML5. In any case, it's sometimes nice to have an abstraction to control the aspects of dealing with sound in the browser whatever the HTML support, and for controlling fallbacks for older browsers. There are a few of these libraries out there—for instance, Buzz.[8] One that I've used in the past, and that I use in our code examples for this chapter, is called howler.js.[9] We'll see some Howler code in the next section.

---

5.   http://caniuse.com/#search=web%20audio
6.   http://www.opus-codec.org
7.   http://www.w3schools.com/html/html5_audio.asp
8.   http://buzz.jaysalvat.com
9.   http://goldfirestudios.com/blog/104/howler.js-Modern-Web-Audio-JavaScript-Library

Dealing with Cross-Site-Scripting Issues

One other headache to deal with when working with sound is that if you plan on using JavaScript and Ajax requests to load the sound files, you must load your site through a web server even when developing locally; you can't simply work from a local directory using the file:// scheme. This is because of restrictions browsers place on loading code to avoid cross-site scripting attacks.[10]

Running the Example Code

Howler.js loads sounds using JavaScript, so when running the example code for this chapter you can't simply open index.html in your browser. If you have a local web server, you can copy the code into its root or set up a site and run the code from there. If you have Python installed on your system, you can open the sample code directory in your shell, change the directory to web/public, and type this:

```
python -m SimpleHTTPServer
```

That will open up a web server with the root at that directory at port 8000. Open http://localhost:8000/index.html in your browser, and you should be able to start the game. Now let's have a look at the example code and see how the sounds are triggered.

Integrating with the Web Game

Before we start looking at code, remember that the most important goal of this chapter is not to cover the web code, but rather to show this game as an example of a case where you might find yourself wanting to export static sound files from Pure Data. If you learned something from the patch for this example but don't particularly care about web code, then this chapter has already been a success.

Also, the example code for this game uses a lot of different frameworks or libraries, such as jQuery, gameQuery, and howler.js. If these frameworks are new to you, or you find yourself confused by the code, take some time to look into these, and come back later. But if you're familiar with web code or you're brave enough to ignore what you don't understand for the time being, let's forge ahead and look at how to integrate these static files exported from Pure Data into a web game.

---

10. http://en.wikipedia.org/wiki/Cross-site_scripting

The Project's Structure

Inside the downloadable code for this book, there's a directory called web. In that directory's root is the Sinatra configuration file, srv.rb, and a directory called public. The project files we'll review now are all stored in that directory.

There are a number of graphics files, and a few sound files:

- thruster.wav
- thruster.mp3
- thruster.ogg
- sounds.wav
- sounds.mp3
- sounds.ogg

For each effect there's one OGG, one MP3, and one WAV file. I opened the WAV files we recorded from Pd in my favorite editor, Audition, and saved them as the other formats. You can do this in a number of ways, including with free sound editors like Audacity.

The thruster sounds are, of course, for the thruster effect we exported, but the other files, called sounds, are the rest of the effects combined into one file. We'll look at why that's the case in a second.

Next, there are a few HTML and JavaScript files, including these:

- index.html
- actors.js
- game.js

The index.html file is where all the important JavaScript files are included and the initial screen set up, but it doesn't contain anything else of interest to us right now. The actors.js file contains JavaScript objects for the player's ship and the asteroids. Poke around in there if you like, but game.js is where all of the game logic resides. The following code snippets all come from there.

Setting Up the Sounds

First let's look at the top of the file where we initialize the howler.js objects we'll use to trigger the sounds. This takes place in a function called initialize-Sound, shown here.

```
web/public/game.js
var sounds, thruster;
function initializeSound() {
  thruster = new Howl({
    urls: ['thruster.wav','thruster.mp3','thruster.ogg'],
```

```
      loop: true,
  });
  sounds = new Howl({
    urls: ['sounds.wav','sounds.mp3','sounds.ogg'],
    sprite: {
      asteroid: [0, 340],
      ship: [351, 1040],
      missile: [1400, 289]
    }
  });
  thruster.play();
  fireThruster(false);
};
```

Two variables, sounds and thruster, are declared here and then each is initialized to be a new Howl, which is the type of howler.js sound object.

Each Howl takes an initialization object. Both thruster and sounds have a urls property in their initialization objects, which is an array containing all the formats for the sound. Howler.js will figure out the right one to download and use based on which browser is accessing the page and whether it supports HTML5 audio and other considerations.

The thruster object is also configured with loop: true. This is because we can't predict how long the player will hold down a movement key, so the thruster sound has to loop over and over until the user lets the key up. This presents its own challenges, but for now note that we call thruster.play() right away, followed by fireThruster(false). This means the thruster sound is playing all the time, and the convenience method fireThruster then mutes it. We'll look at this more in a second when we talk about the thruster sound.

Now let's look at the sounds object's configuration. Notice that there's a configuration option called sprite here. This contains an object with keys for the other types of sounds we produced from the patch, each with values of a list of numbers. Howler.js is using the concept of a sound sprite here.

Sound Sprites

A *sprite sheet* is a production graphic concept where many different images are put together into one image, laid out in a grid. There are a few advantages to this approach: related graphics, such as those being used as frames in an animation, can be kept in one place, and the browser needs to download only one file. This final point is interesting, because sometimes the limiting factor for a browser to quickly download and display a web page is not the size of the assets it has to download, but the number of connections it needs to have open to download all the assets. Often a medium-sized sprite sheet can be downloaded faster than the many smaller images contained in the sheet if they were separate files.

The idea can carry over to sounds, too, and howler.js gives us the ability to combine a set of sounds into one file and define sprites in that file by telling it the starting point and duration. That's what we've done in this configuration of the sounds object. Now let's see how and where we trigger each sound.

Triggering the Sounds

To play a sprite contained in a howler.js sound sprite, we simply pass an argument of the sprite name to any howler.js object's play method. A good first example is the asteroid-explosion sound.

The Asteroid Explosion

The code for dealing with missiles fired by the spaceship takes place inside of a callback function that executes very frequently. Inside of that function a number of game logic functions are performed on whatever game objects are on the screen. The following is the logic for the missiles.

```
web/public/game.js
$(".playerMissiles").each(function(){
  var posx = $(this).x();
  if(posx > PLAYGROUND_WIDTH){
    $(this).remove();
    return;
  }
  $(this).x(MISSILE_SPEED, true);
  var groupName = ".asteroid,." + $.gQ.groupCssClass;
  var collided = $(this).collision(groupName);
  if(collided.length > 0){
    collided.each(function(){
      $(this).setAnimation(asteroid["explode"],function(node){$(node).remove();});
      $(this).removeClass("asteroid");
    });
    $(this).setAnimation(missile["explode"],function(node){$(node).remove();});
    $(this).removeClass("playerMissiles");
    sounds.play('asteroid');
  }
});
```

First, a jQuery lookup is done with the selector .playerMissiles to find all the missiles on the screen. The code updates the position of each of these, removing them if they're offscreen, and then uses the collision function to look for any asteroids each missile has collided with.

It iterates through each of these asteroids, if any, tells them to play their explosion animation and remove themselves, and then tells the current missile to play *its* explosion animation and remove *itself*. Finally, at the end of all this, if a collision occurred we play the asteroid sprite of the sounds object, which is the low *boom* sound from the patch.

The Ship Collision

The code for determining if an asteroid has hit the ship is very similar.

web/public/game.js
```
$(".asteroid").each(function(){
  this.asteroid.update();
  var posx = $(this).x();
  if((posx + 100) < 0){
    $(this).remove();
    return;
  }
  var collided = $(this).collision("#playerBody,."+$.gQ.groupCssClass);
  if(collided.length > 0){
    $(this).setAnimation(asteroid["explode"], function(node){$(node).remove();});
    $(this).removeClass("asteroid");
    if($("#player")[0].player.damage()){
      explodePlayer($("#player"));
    }
    sounds.play('ship');
  }
});
```

In the context of iterating through each asteroid object jQuery finds on the screen, offscreen asteroids are cleaned up, and any asteroids that have collided with the ship cause the ship damage. If two asteroids have hit the ship so far, this third collision should cause the ship's damage method to return true, and the helper method explodePlayer is called to animate the ship's explosion off of the screen. If the ship has exploded three times, the game is over.

Finally, if any collision has occurred, no matter if it causes an explosion, we play the sounds object's ship sound sprite.

The Missile-Firing Sound

The missile-firing sound is triggered in some key-handling code, shown here.

web/public/game.js
```
$(document).keydown(function(e){
  if(!gameOver && !playerHit){
    switch(e.keyCode){
    case 75:
    case 32:
      var playerposx = $("#player").x();
```

```
var playerposy = $("#player").y();
var name = "playerMissile_"+Math.ceil(Math.random()*1000);
var options = {
  animation: missile["player"],
  posx: playerposx + 90,
  posy: playerposy + 14,
  width: 22, height: 10
};
$("#playerMissileLayer").addSprite(name,options);
$("#"+name).addClass("playerMissiles");
sounds.play('missile');
break;
```

If the game isn't over and the player's ship isn't hit and exploding off of the screen, a switch statement decides if a key we care about has been pressed. If the key code is 75 or 32, a k key or spacebar, then a missile is added to the screen at the ship's current position. This takes a few lines to accomplish, but at the end we trigger the missile sprite of the sounds object.

The Thruster Sound

Now let's look back to the thruster sound. In the initialization code, we finished by playing the thruster sound and then muting it because we don't want to play a separate thruster sound each time the player presses a movement key. That would sound like a bunch of thruster sounds all at once. Instead we just want to turn one sound on or off depending on if the player is moving the ship.

At first I thought I'd simply play the thruster sound and mute or unmute it, but I ran into a problem with Firefox during testing. The current solution is still not perfect on Firefox, but most of the time it's fairly close to what we want. Again, dealing with browser inconsistencies is the web developer's lot in life.

The following is the code to deal with turning the thruster sound on and off.

web/public/game.js
```
function fireThruster(fire) {
  // thruster.mute seems to not work on Firefox
  if (fire) {
    thruster.volume(1);
  } else {
    thruster.volume(0);
  }
}
```

Continuing the key-down handler code from where the missile-firing keys are caught, here we deal with the movement keys being pressed.

```
web/public/game.js
case 65:
  $("#playerBooster").setAnimation();
  fireThruster(true);
  break;
case 87:
  $("#playerBoostUp").setAnimation(playerAnimation["up"]);
  fireThruster(true);
  break;
case 68:
  $("#playerBooster").setAnimation(playerAnimation["booster"]);
  fireThruster(true);
  break;
case 83:
  $("#playerBoostDown").setAnimation(playerAnimation["down"]);
  fireThruster(true);
  break;
```

And finally, in this key-up handler code, we turn the thruster sound off if the key code matches a movement key.

```
web/public/game.js
  $(document).keyup(function(e){
    if(!gameOver && !playerHit){
      switch(e.keyCode){
        case 65:
          $("#playerBooster").setAnimation(playerAnimation["boost"]);
          fireThruster(false);
          break;
        case 87:
          $("#playerBoostUp").setAnimation();
          fireThruster(false);
          break;
        case 68:
          $("#playerBooster").setAnimation(playerAnimation["boost"]);
          fireThruster(false);
          break;
        case 83:
          $("#playerBoostDown").setAnimation();
          fireThruster(false);
          break;
      }
    }
  });

  initializeSound();
});
```

That covers all of the sounds that the game design called for, integrated into the code and triggered through the howler.js objects.

Things to Think About

Here are a few things to think about with regard to the game's implementation:

I mentioned that I thought about a "red alert" alarm sound for when an asteroid hits the ship. Try creating one. Instead of the falling tones that are in most of the sections of the patch, try a rising tone that goes from one given frequency to another.

Do users always want sound? We've at least not started out the game by automatically making noise, but it'd also be good to add a volume control, or at least a mute button. The howler.js documentation on global methods is a good place to start.

Next Up

Designing sounds for a game is fun. It poses creative challenges and, depending on the delivery platform, some logistical issues to deal with. As this chapter shows, Pure Data is still a handy tool if used only for the sound design, recording, and exporting the final assets for use in other platforms.

In this chapter we covered a lot. We discussed some of the characteristics that define the sound of 8-bit consoles, and then we saw a patch that reproduced some fun, sci-fi sound effects emulating these characteristics. The patch also built on what we've done before with exporting sound files from Pd and showed a more practical example of how to do this.

You saw some of the challenges we face when delivering sound to the web platform, and a few possible solutions to overcome these, courtesy of the howler.js audio library.

Next we'll go through one more project example on native mobile platforms, which allows us to do away with the need to work with static audio files. We'll build an app for Android and iOS and use an excellent library that allows us to embed Pd directly in the apps.

CHAPTER 9

Integrating Dynamic Sounds into a Native App

And now comes the *pièce de résistance*, when we're able to take everything we've covered so far and put it into an app, allowing us to embed Pure Data and have a completely dynamic audio engine in our mobile apps. We'll go over the design of an app for both Android and iOS, including the goals and design of a Pd patch to include sound. Then we'll use a library called libpd to integrate the patch into the apps. We'll finish by going over the code in each platform, integrating Pd into the apps.

When you are done with this chapter, you will have

- Designed the sound for a set of native apps
- Built a configurable, dynamic audio patch for production use
- Integrated Pd into native app code

Now let's get started by looking at the design of the app we'll integrate Pd into.

Designing the App

We have a few goals in creating this app. At the highest level, of course, we want to get some experience integrating Pd into native apps. We also want to think a bit about good sound design for an app.

One of the best examples of good sound design from a recent app is Clear,[1] a task-management app for iOS, with sound design by Josh Mobley.[2] It follows good sound-design principles. It doesn't have a sound effect for everything

---

1. http://www.realmacsoftware.com/clear/
2. http://joshmobley.com/

just for the sake of having a sound effect; most of the effects serve a purpose. It uses sound effects to do more than just tell users when they've clicked a button or interacted with the app; it uses sound to emotionally engage users and to encourage them to use the app to achieve their goals.

With that in mind, we'll build a task-management app for iOS and Android and incorporate sounds in the same manner as the Clear app. In our app, though, we won't use static samples—we'll use Pd to create a patch that will allow us to express different audio cues and effects dynamically with certain configuration left up to the hosting app.

Creating Requirements for a Task-Management App

Tasks-management apps are a dime a dozen, or maybe more like $12 a dozen, taking inflation and current app-store prices into account. That's OK, because since it's a simple problem it will let us focus on the sound-design aspect.

Let's take a look at the features the app will implement. Here are the main ones:

- Create a new task
- Edit a task description
- Mark a task as complete and incomplete
- Remove completed tasks

This is a good enough feature set to work for a simple task-management app, and it gives us a solid base of user stories to add some audio cues.

Designing the Sound Effects

As mentioned, the effects for even this simple app should have some sort of goal beyond just adding sound. Here is a good set of goals:

- Include audible feedback for some actions
- Encourage users as they complete tasks
- Reward users as they remove completed tasks

The audio should layer on top of the application's features while not being essential to the app's functions. It's easy enough to double up on the user-interface (UI) control interaction and provide a sound for something the user already has visual confirmation of. We'll do that to give a little extra feedback that the user has added a new task, but we can use sound to add an extra dimension, too.

The Clear app plays a set of pleasantly rising tones as the user completes tasks, and we'll take the same approach. This technique works psychologically, defining a rising audible progression that makes the user anticipate the completion of the next task and adds a little incentive to complete tasks. When the user clears off a set of completed tasks we'll play another pleasant tone that signifies completion and acts as a small reward, which encourages the user to come back and complete more tasks.

The progression we'll use is called a *major scale*, and the tones we'll use are called *major thirds*. A scale is a certain set of musical notes that go up from a given note, follow a certain pattern, and end up on an *octave* above that *root* note. Any musical scale starts on a note, called the root, and goes up by defined steps. The octave is a note that is double the frequency of a given note, and is where the scale starts the pattern again. A third is the third step in the scale, and a major third is the third step in the major scale. We'll use the major scale because it's generally perceived as having a happy or positive quality.

Let's take a quick look at the visual design for the app before we dive in and start creating a patch to accomplish our sound-design goals. These are the audio cues we'll add:

- A short tone when a task is created
- A major third when a task is completed, stepping up the major scale for each task completed after that
- Two octaves of major thirds in short sequence when the completed tasks are cleared
- The creation tone for when a task is marked incomplete

Each version of the app has a minimal UI to allow us to get the sounds into the app. We could add more polish or features, but this is a good enough set to focus on the sound design and integrate Pure Data.

Android Version

The Android version follows the standard pattern for an Android app, with an action bar with the main controls, a menu for extra controls, and a list view for the tasks. As Figure 32, *Adroid App New Task*, on page 142 shows, tapping the New Task menu item creates a new task, which can be edited by tapping the task-list item, which will play a short audio cue.

Once a task is created, it has a check box in the list, which will mark the task as complete. (See Figure 33, *Adroid App Complete Task*, on page 142.) The user can tap the check box again to mark the task as incomplete.

Now let's look at the iOS version.

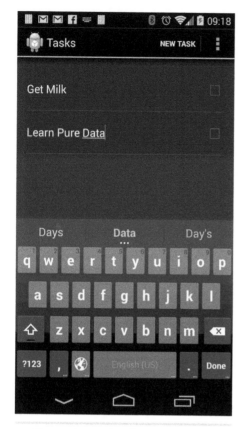

Figure 32—Adroid App New Task   **Figure 33—Adroid App Complete Task**

iOS Version

The iOS version of the app follows the interaction of the Clear app more closely. Users are presented with a simple screen (shown in Figure 34, *Adroid iOS App Add Task*, on page 143), which they can pull down on to create a new task.

Tasks are presented in a list, as shown in Figure 35, *Adroid iOS App New Task*, on page 143, and the user can tap into a task to edit it.

To complete the task, the user swipes the task item to the right, which animates that task into the bottom of the list, where the completed tasks are grouped. To clear all completed tasks, the user taps a completed task and is presented with a confirmation. If the user confirms, the completed tasks are removed.

Now that we have a clear idea what we're working toward, let's take a look at the Pd patch we'll create to accomplish these sound-design goals.

Figure 35—Adroid iOS App New Task

Figure 34—Adroid iOS App Add Task

Designing and Building the Patch

Keeping in mind the goals we have for the app, let's design a patch that will do what we need.

Design Goals

We've decided we'll use musical tones to create the feel we want for the app, including encouraging the user to complete tasks. Now let's consider the type of sound we want. There's nothing scientific here; in other words, we're not trying to make a sound effect to match a real-world sound. As with most musical things, this is largely a matter of taste. However, we want to place a few constraints on the design, given the fact that the target is a mobile device.

- We want a clear, short sound.
- It must cut through background noise well on mobile speakers.
- We want an upbeat, positive feel.

For those reasons, let's use a bell-like tone for the sound, one that has a bit of an attack and is harmonically rich. This will help it cut through better on small speakers, and will make a nice, interesting sound.

Designing the Sound

There are a number of ways to make a bell-like sound with synthesis. We did much the same thing when we modeled the wineglass strike, adding the exact frequencies we wanted using additive synthesis. We could also use filters to cut away or emphasize the frequencies we wanted: subtractive synthesis. Or we could capture the sound we wanted and play back the sample or use it to build a wavetable for wavetable synthesis.

There's one more type of synthesis we'll look at now, though, which perfectly fits the criteria we have. It's called *frequency-modulation synthesis*. It was discovered in the '70s and developed in the '80s as a way to make complex timbres with only a few oscillators, which made it great for the low-powered musical synthesizers of the time. It also is characterized by harmonically rich and interesting bell-like or metallic tones.

FM Synthesis

The idea of frequency-modulation (FM) synthesis is to use one oscillator, called the *carrier*, to produce the desired musical note, and another oscillator, called the *modulator*, to change the carrier's frequency. The result is that the output's frequency has a very complex harmonic character for the processing cost of just a few oscillators. Note, however, that oscillators are usually called *operators* when speaking about FM.

Designing Our API

Now let's talk about the patch we want to create. We want it to work inside the native mobile apps and have some controls exposed to them.

Our goal is not only to build a patch to create the tones we want, but to add the configurability to allow the patch to be controlled the way we want from inside the mobile apps we create. I find that it helps to think of this as no different from creating a code module or API that we expect to expose to some other software.

We know we want a way to play a distinct musical tone, so we should have a way to specify the note we want the patch to play. Pd can easily receive MIDI messages, which you'll recall are note and control messages passed to and from digital musical instruments. In this case, since we don't plan on using a musical instrument to control Pd, we don't really need to use a MIDI interface. It's still convenient to use MIDI to specify the notes we want to play, however, so we'll have a way for the patch to receive a MIDI note number and play it.

As an extra feature thrown in at the last minute (as software developers we wouldn't feel at home without a few of these), let's also expose a few other settings that allow the native apps to slightly change the characteristics of the tone so the Android app can sound distinct from the iOS app.

The Task-App Patch

Now we'll walk through the patch. Follow along if you like by creating a patch in a working directory. To ship the patch with the app, we'll need to copy the patch into the Android and iOS app directories. For iOS it's enough to include the patches in the main bundle. For Android, the patches need to be zipped up and placed in the raw resources directory. More about Android later, when we go over the Android code.

We'll start at the top of the patch, and then work through some abstractions. You can find the final patch in the code download.

The Top-Level Patch

To begin with, look at the top-level patch in Figure 36, *The Top-Level Task-App Patch*, on page 146. There are two "voices," or subpatches, named bellvoice~ with a number of receive objects connected to inlets and the same arguments sending output into throw~ out objects. The catch~ out object is connected to a dac~.

We have two separate bellvoice~ subpatches so we can play two notes that sound at the same time. In most musical synthesizers this capability is referred to as the number of *voices* it supports. Notice that the top voice has an r midinote1 and the bottom has an r midinote2 connected to the first inlets. We'll send a message to each of these MIDI note receivers when we want the two separate notes to sound. We have here, then, an FM synthesizer with two voices.

The rest of the receive messages make up the rest of the API for the patch. There are two sets of tune and depth messages because each bellvoice~ has two operator pairs, for a little extra character to the sound. The tuning value isn't the frequency we want the modulators to have; it's actually a multiplier of the note frequency sent to the carrier. So, the modulation frequency is always expressed in terms of the carrier frequency in this design.

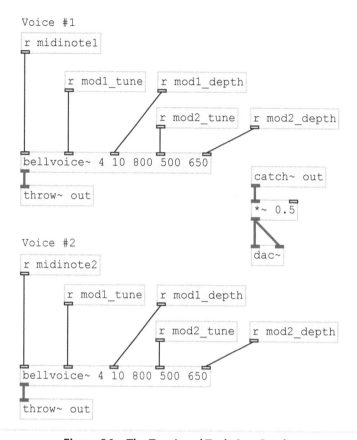

Figure 36—The Top-Level Task-App Patch

The MIDI note messages control the frequency of the carrier operators, and the other messages control the characteristics of the modulators. The messages are as follows:

- mod1_tune: tuning for the first modulator, as a ratio of the first carrier frequency

- mod2_tune: tuning for the second modulator, as a ratio of the second carrier frequency

- mod1_depth: the depth of the first modulator

- mod2_depth: the depth of the second modulator

Going back to what we covered about FM, the tuning of the modulators will set their frequency, which is the rate at which the carriers change their frequency. The depth messages will control the modulators' amplitude. One way of looking at this is how much the modulators modify the carrier, so I've used the word *depth*. In FM this is often referred to as the *index* of the modulator, so keep that in mind. The specs for FM are sometimes a little academic sounding.

Those are the messages we'll use as our API for the mobile apps to interact with. Five other arguments are passed to the bellvoice~ subpatch, however, so let's talk about those, and the subpatch itself, next.

A Single Voice

The following figure shows the image for a single voice subpatch.

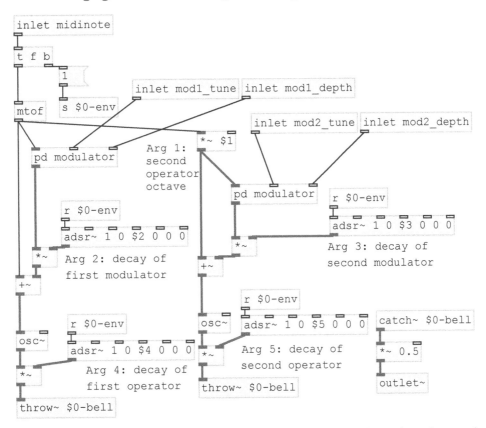

This subpatch is saved in the same directory as the top-level patch and named bellvoice~.pd. The patch is fairly dense, but conceptually there are two operator pairs. Each operator pair is composed of a carrier (which is simply an osc~) and a modulator (which is a subwindow, pd modulator).

The modulators each accept the tune and depth inlets and are connected to +~ objects. In this way they send out a signal of a certain frequency and amplitude, depending on their settings, which is summed with another frequency, thus modulating it. That frequency is sent to the osc~ objects, which send their signals to the outlet using a throw-and-catch pattern.

The original frequency is derived from the inlet midinote fed into the mtof object, which turns a MIDI note number into a frequency float value. This is how we specify the tone of the note each voice plays.

Finally, both the carriers and modulators are multiplied by asdr envelopes so that they can each be given some timing control. Giving a modulator a longer or shorter decay time than its carrier will change the character of the sound you hear while the note plays. These envelopes are each labeled for clarity because the decay values are all dollar-sign variables, taken from the arguments to the bellvoice~ subpatch on the parent patch.

Notice that the adsr~ subpatch is not, as it was in other examples, inside of the BuildingControls directory, but rather in the same directory as this set of patches. This is because we must copy all the patches we need into the native app project to build them into the app.

The last thing to note is that the second operator pair on the right of the patch has its signal multiplied by the value in $1 first. This allows us to layer the second pair some octaves above or below the first, which produces a more interesting sound. This isn't a necessary ingredient of the patch; I just thought it sounded better. While you design patches, don't forget to play and experiment!

A Modulator Operator

Finally, the subwindow content of each pd modulator looks like the following image—first, an inlet~ for the frequency being sent to the carrier; second, the tuning multiplier inlet; and then the depth inlet.

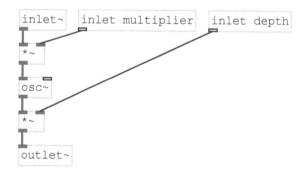

The incoming frequency is multiplied by the tuning value, fed into the osc~, and then multiplied by the depth setting, controlling the amplitude, or index of the modulator in FM terms.

Testing the Patch

This is a fairly dense patch, and after all this talk about how FM works, I'm sure you're eager to hear how it sounds. A nice technique, which we've already used in previous patches, is to add some message boxes to send the values to each receive object. In this way we can fine-tune values we expect to be coming from the mobile apps directly in the patch, and then use those values to configure our native code.

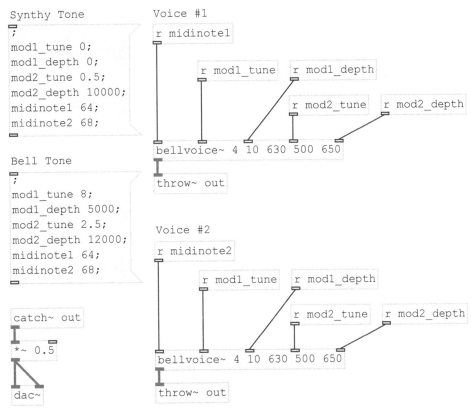

The preceding image shows the final top-level patch with two message boxes that hold some settings I've chosen to produce two interesting sounds from the patch we've built. Both use two MIDI notes, 64 and 68, to create a major third, sort of a happy sound.

Two Different Tones

Beyond that, the settings are slightly different. The message box labeled Bell Tone sounds quite like a bell, with a harder initial attack and a metallic quality. Synthy Tone sounds slightly bell-like, but has a more synthesizer-like quality. I experimented with the values and landed on these as the tones I wanted to use in each app—the synthy tone in the Android app and the bell tone in the iOS app.

The Hard-Coded Settings

Let's go over the settings in the message boxes and what they're doing. But first, to understand them we need to understand the non-configurable settings in the arguments to bellvoice~:

- Second operator octave multiplier: 4
- Modulator 1 decay: 10ms
- Modulator 2 decay: 630ms
- Carrier 1 decay: 500ms
- Carrier 2 decay: 650ms

These settings mean the second carrier group is set to be four octaves above the first, adding some depth to the sound. The first modulator affects only the first carrier for a quick burst of 10ms. This is great for making the initial bell attack sound, simulating metal being struck by some object, like a bell-striker.

The second modulator is set to be a little faster to decay than the second carrier, so the sound changes character subtly over time. The second carrier, which is the one four octaves higher than the first, is set to decay 150ms after the first carrier, so that the higher-pitched sound hangs around just a little longer than the lower, which mimics the effect of a bell.

The Bell-Tone Settings

Now have a look at the settings in the message boxes. Notice that the bell sound has a tune and a depth set for the first modulator. It doesn't matter much what the tuning is, and the depth affects the sound only a little because the decay for the first modulator is a very fast 10ms to simulate the attack portion of striking a bell.

The tuning of the second modulator is 2.5, which produces an interesting metallic quality in the sound from the second carrier because it's not an even multiple of the frequency. The depth of this modulator is set rather high as well, to heighten the effect.

The Synth-Tone Settings

For the synth tone, we set the fast-attacking first modulator to 0, which removes the initial percussive sound. We also turn down the second modulator's tuning, so it's still not an even multiple, but it's closer in pitch than the bell tone. Its depth is also slightly lower. These settings were derived mostly by experimenting with the bell settings once I had what I liked there.

Now we have a solid FM synthesizer patch with configurable settings that we can control from our native apps. Let's dive right into seeing how we can accomplish that.

> ### Things to Think About
>
> Play around with some different settings, and explore how FM synthesis works and sounds.
>
> This should be an easy one: Why are each of the catch~ objects in the top level and submodule patches preceded by a *~ 0.5?
>
> What if we wanted to expose more settings to the native apps, such as the decay rates of the carriers? How would we do that?

Introducing libpd

Now we have a Pd patch and we have the design for the native apps, but we need a bridge between the two. Let's talk about libpd,[3] the excellent library that allows us to embed Pd right into our apps, with interfaces to handle two-way communication.

About libpd

Libpd started out as an experiment by Peter Brinkmann to get Pure Data working on Android, but has developed into what we have today: the ability to run an instance of Pd in an Android or iOS app. We'll use libpd in both of our implementations of the task app. Refer to the libpd website for specific integration instructions for each platform, but let's first talk about a few things to keep in mind when building patches meant to be used with libpd.

Third-Party Extensions

When you integrate libpd into your project, the Pd instance that's running is Pd Vanilla, not Pd Extended. This makes it important to design your patches mindfully using third-party extensions and make sure you test on the device.

---

3.   http://libpd.cc/

You can use even extensions written in C; for more information see the documentation at the libpd documentation site.[4]

If you do use third-party extensions, make sure you take note of the license they're released under and understand the legal restrictions. If an extension uses the GPL,[5] the LGPL,[6] or another license that puts conditions on the distribution of the software using it, you will have to navigate the legal issues before putting your app on any of the marketplaces.

We've gotten accustomed to using one extension that ships with Pd Extended, output~. That extension won't be available to the Pd instance running under libpd, and it doesn't really have any benefit to a patch we'll control with native code, so in the patch for these apps we've used a dac~ instead.

Now let's dig into the code and see how to make the magic happen.

A Note About the Code

This book is primarily about Pure Data, but for a mostly programmer audience. I implemented both Android and iOS apps so that programmers with either Java or Objective-C under their belts could benefit from reading the code, but we won't go over the basics of those platforms.

If you are familiar with a given platform, this should be fine for you. If you're curious about the implementation for a particular platform, dive in and see how it works, but in the following section we'll focus only on the parts that have to do with interfacing with libpd.

Integrating with the Native Apps

Now we have a patch that sounds the way we want and that we can configure a bit to our taste, and it presents an API we can work with in the native apps. Let's look at the app code for the Android and iOS platforms side by side, and see how we use libpd to control the patch.

If you're not familiar with one of the platforms, just ignore that code or compare and get a sense for similarities and differences. In pretty much all the cases used in these apps, the code is strikingly similar.

---

4.   http://libpd.cc/documentation/
5.   http://www.gnu.org/copyleft/gpl.html
6.   https://www.gnu.org/licenses/lgpl.html

Setting Up Pd

Let's start with Android. The structure of the Android app is pretty standard, with one main activity, TasksActivity. Since there are a few different user interactions in this activity and the list item, I've consolidated the libpd setup and interaction into a singleton, PdInterface. When we want a sound to occur anywhere in the Android view code, we simply tell this singleton to play the sound we want. In the life-cycle method onCreate, in the following code, you can see where we initialize PdInterface, passing in a reference to a context as the activity.

```
android/TasksProject/Tasks/src/main/java/com/programmingsound/tasks/TasksActivity.java
@Override protected void onCreate(Bundle savedInstanceState) {
  super.onCreate(savedInstanceState);
  setContentView(R.layout.activity_tasks);
  listView = (ListView)findViewById(R.id.task_list);
  PdInterface.getInstance().initialize(this);
  loadTasks();
}
```

Inside PdInterface initialization methods, the main libpd setup takes place.

```
android/TasksProject/Tasks/src/main/java/com/programmingsound/tasks/audio/PdInterface.java
private void initializePd() throws IOException {
  AudioParameters.init(context);
  int sampleRate = Math.max(MIN_SAMPLE_RATE, AudioParameters.suggestSampleRate());
  int outChannels = AudioParameters.suggestOutputChannels();
  PdAudio.initAudio(sampleRate, 0, outChannels, 1, true);
```

Android developers are intimately familiar with dealing with different devices and capabilities in their code. Libpd makes it easy to configure itself by using AudioParameters to suggest a few things based on the current system, such as sample rate and number of output channels.

We initialize the parameters singleton with our context, decide on a sample rate and number of output channels, and initialize another singleton called PdAudio. The minimum sample rate we'll accept, 44100, is configured in the constant MIN_SAMPLE_RATE.

Next we deal with unpacking the patch out of the app's resources, shown in the following code. The standard way of shipping supporting files like the Pd patches with an Android app is to put them in the raw resource directory.

Android doesn't like anything strange in the filenames of these resource files, like a .zip or even a tilde, as we have in bellvoice~.pd, so when we ship the patch with the app we compress all of the patches we use—task_tones.pd, bellvoice~.pd, and adsr~.pd—into a ZIP file and name it patches in the raw resources directory. The following code unpacks this into the content's files directory.

```
android/TasksProject/Tasks/src/main/java/com/programmingsound/tasks/audio/PdInterface.java
File dir = context.getFilesDir();
File patchFile = new File(dir, "task_tones.pd");
InputStream patchStream = context.getResources().openRawResource(R.raw.patches);
IoUtils.extractZipResource(patchStream, dir, true);
```

Notice that libpd comes with a convenient helper class, IoUtils, to take care of the unzipping operation.

After the patches are unzipped, we use the File object created in the preceding code and tell libpd to load it.

```
android/TasksProject/Tasks/src/main/java/com/programmingsound/tasks/audio/PdInterface.java
    PdBase.openPatch(patchFile.getAbsolutePath());
    PdBase.sendFloat(MODULATOR_1_TUNE, 0);
    PdBase.sendFloat(MODULATOR_1_DEPTH, 0);
    PdBase.sendFloat(MODULATOR_2_TUNE, 0.5f);
    PdBase.sendFloat(MODULATOR_2_DEPTH, 10000);
}
```

Then we configure the patch with the same values as were in the Synthy Tone message box we used to test inside the patch. Notice that we use constants such as MODULATOR_1_TUNE instead of string literals like mod1_tune, so we don't have magic strings laying all over the code referencing Pd objects and messages in case these change.

To communicate with Pd, we use the appropriate PdBase class method. In this case, we want to send some messages, just like we did from the message box in the patch. They're float values, so we use sendFloat. Simple, isn't it?

Now Pd is initialized and configured the way we want for the current Android system. Let's look at the iOS setup. The app's structure is also idiomatic iOS. I've consolidated the libpd setup and interaction into a single PdInterface class with a reference to an instance on the AppDelegate. When we want a sound to occur anywhere in the view code, we simply tell this instance to play the sound we want.

In the life-cycle method application:didFinishLaunchingWithOptions: in the following code, you can see where we initialize a PdAudioController with a sample rate and some options.

```
ios/Tasks/Tasks/AppDelegate.m
self.pdAudioController = [PdAudioController new];

PdAudioStatus status = [self.pdAudioController configureAmbientWithSampleRate:44100
                                              numberChannels:2
                                              mixingEnabled:YES];
```

```
if (status == PdAudioError) {
    // handle audio initialization error - probably by doing nothing.
} else {
    self.pdInterface = [PdInterface new];
}
```

This object is responsible for initializing the Apple audio libraries the way we want. Here we've told libpd to initialize with ambient playback allowing mixing, which means that we should not shut off any other audio playing in the background as we open this app.

Next, if that initialization works out, we create an instance of PdInterface. The following is its initialization code:

ios/Tasks/Tasks/PdInterface.m
```
- (id) init {
    self = [super init];
    if (self) {
        scaleDegrees = @[@0, @2, @4, @5, @7, @9, @11, @12, @14, @16];
        self.pdDispatcher = [[PdDispatcher alloc] init];
        [PdBase setDelegate:self.pdDispatcher];
        patch = [PdBase openFile:@"task_tones.pd"
                            path:[[NSBundle mainBundle] resourcePath]];
        // Initialize the FM operator settings. Could be read from config
        [PdBase sendFloat:8. toReceiver:kModulator1Tune];
        [PdBase sendFloat:5000. toReceiver:kModulator1Depth];
        [PdBase sendFloat:2.5 toReceiver:kModulator2Depth];
        [PdBase sendFloat:12000. toReceiver:kModulator2Tune];
        [self resetScaleDegree];
    }
    return self;
}
```

Here we store an array of numbers to be used to determine position in a scale, which we'll talk about shortly. Then we initialize a PdDispatcher and add it as a delegate to PdBase. Next we load up our patch file through PdBase. Unlike with Android, we don't need to zip up the patches; we can just include them in the main bundle.

Finally, we send a series of messages to Pd to configure everything the way we did using the Bell Tone message box in the patch. Again, notice that we use constants such as kModulator1Tune instead of string literals like mod1_tune, so we don't have magic strings laying all over the code. Then it's as simple as calling sendFloat:toReceiver: on PdBase, and we're communicating with Pd from iOS!

Now that we're initialized and configured, let's see how we can make all the sounds we want to make.

Wiring Up the New Task Cue

When the user creates a new task, we want to play a short tone. Let's see how we do that. To create a task, the user pulls down the UICollectionView on iOS or taps New Task on Android; that calls back to the following methods.

android/TasksProject/Tasks/src/main/java/com/programmingsound/tasks/TasksActivity.java
```
private void createNewTask() {
  Uri tasksUri = TasksContentProvider.CONTENT_URI;
  ContentValues values = new ContentValues(2);
  values.put(TasksDBHelper.TASK_TITLE_COLUMN, "New Task");
  values.put(TasksDBHelper.TASK_COMPLETE_COLUMN, 0);
  getContentResolver().insert(tasksUri, values);
  PdInterface.getInstance().playNewTaskCue();
}
```

ios/Tasks/Tasks/TaskListViewController.m
```
- (void) addNewTask {
    [self.refreshControl endRefreshing];
    NSManagedObjectContext *moc = [AppDelegate sharedInstance].managedObjectContext;
    NSEntityDescription *desc = [NSEntityDescription entityForName:@"Tasks"
                                      inManagedObjectContext:moc];
    NSManagedObject *task = [[NSManagedObject alloc] initWithEntity:desc
                                      insertIntoManagedObjectContext:moc];
    [task setValue:@"New Task" forKey:@"title"];
    [task setValue:[NSDate date] forKey:@"createdAt"];
    [self.pdInterface playTaskCreatedCue];
    [self.collectionView reloadData];
}
```

There we do the work of inserting the task into the tasks database, but on the second line of the methods we tell our PdInterface to play the new task audio cue.

android/TasksProject/Tasks/src/main/java/com/programmingsound/tasks/audio/PdInterface.java
```
public void playNewTaskCue() {
  PdBase.sendFloat(MIDINOTE1, ROOT_MIDI_NOTE);
  resetScaleDegree();
}
```

ios/Tasks/Tasks/PdInterface.m
```
- (void) playTaskCreatedCue {
    [PdBase sendFloat:kRootMidiNote toReceiver:kMidiNote1];
    [self resetScaleDegree];
}
```

These methods in the Pd wrapper code simply send the ROOT_MIDI_NOTE or kRootMidiNote constants, set to 60, or the note C, to the "midinote1" receiver in the patch. Then we call a method named resetScaleDegree. Let's talk about what that does now.

Playing a Scale as Users Complete Tasks

When users complete a task, we want to play a two-note musical sequence as a cue and a small reward. But if they complete a sequence of tasks, we want to step up a musical sequence to create a sense of progression and encouragement to complete more tasks.

The two-note musical sequence will be a *major third*, and the upward-moving musical sequence will be a *major scale*. If you're a musician, you'll know what I mean right away, but let me briefly explain the musical theory here. A scale is a formula of musical steps that can be started at any note. Once the note is chosen, the formula reveals which notes are at which position. A scale stops just before the octave, where it wraps around and starts over at the first step.

The major-scale formula is as follows: root, two steps, two steps, one step, two steps, two steps, two steps, one step. We encode that into an array of integers as a constant on PdInterface, storing the absolute distance from the root note in a list, initialized in init on iOS and in the code here on Android.

android/TasksProject/Tasks/src/main/java/com/programmingsound/tasks/audio/PdInterface.java
```
private static final int[] scaleDegrees       = {
  0, 2, 4, 5, 7, 9, 11, 12, 14, 16
};
private static final int ROOT_MIDI_NOTE       = 60;
```

The major third is the third position of the major scale. Since we want to play a major third each time the user completes a task, and then step up the scale and play another major third, we just need to keep track of where we are in the scale as we go and then step up until we reach the end of the scale. Another variable will keep track of which octave we're on. The code to do that is in incrementScaleDegree. Here are the Android and iOS versions.

android/TasksProject/Tasks/src/main/java/com/programmingsound/tasks/audio/PdInterface.java
```
protected void incrementScaleDegree() {
  currentDegree++;
  if (currentDegree == 7) {
    currentDegree = 0;
    currentOctave++;
  }
  int rootDegree = scaleDegrees[currentDegree];
  int nextDegree = scaleDegrees[currentDegree + 2];
  currentRootNote = ROOT_MIDI_NOTE + (12 * currentOctave) + rootDegree;
  currentSecondNote = ROOT_MIDI_NOTE + (12 * currentOctave) + nextDegree;
}
```

ios/Tasks/Tasks/PdInterface.m

```
- (void) incrementScaleDegree {
    currentDegree = currentDegree + 1;
    if (currentDegree == 7) {
        currentDegree = 0;
        currentOctave = currentOctave + 1;
    }
    int rootDegree = [scaleDegrees[currentDegree] integerValue];
    int nextDegree = [scaleDegrees[currentDegree + 2] integerValue];
    currentRootNote = kRootMidiNote + (12 * currentOctave) + rootDegree;
    currentSecondNote = kRootMidiNote + (12 * currentOctave) + nextDegree;
}
```

After the user breaks the sequence by either creating a new task or clearing all the completed tasks, we need to reset back to the beginning of the scale, which is what resetScaleDegree does, and why we call it from the new task cue methods.

Now that we know how we'll step up the scale, let's look at completing tasks.

Wiring Up the Complete Task Cue

When users complete a task, we want to play the task-completion cue, but if they uncomplete a complete task, let's play the new-task cue.

android/TasksProject/Tasks/src/main/java/com/programmingsound/tasks/view/TaskListItem.java

```
@Override public void onCheckedChanged(CompoundButton buttonView,
                                       boolean isChecked) {
    if (isChecked) {
        PdInterface.getInstance().playCompleteTaskCue();
    } else {
        PdInterface.getInstance().playNewTaskCue();
    }
    updateTask();
}
```

ios/Tasks/Tasks/TaskCell.m

```
- (void) playToneForGesture {
    if (completed) {
        [self.pdInterface playTaskCreatedCue];
    } else {
        [self.pdInterface playTaskCompletionCue];
    }
}
```

To play the note sequence of a completed task, we need a way to play the first note, and then wait a small amount of time before playing the next note. The solution we'll use is not scalable, so it wouldn't work for a musical app that needed to play a sequence of notes according to some stored representation of the music. However, it is completely acceptable for the small sequence we want to play. The Android solution looks like this.

android/TasksProject/Tasks/src/main/java/com/programmingsound/tasks/audio/PdInterface.java

```java
public void playCompleteTaskCue() {
  final float third = currentSecondNote;
  PdBase.sendFloat(MIDINOTE1, currentRootNote);
  handler.postDelayed(new Runnable() {
    @Override public void run() {
      PdBase.sendFloat(MIDINOTE2, third);
    }
  }, NOTE_DELAY);
  incrementScaleDegree();
}
```

We use a Handler, the way Android wants us to work with simple asynchronous code, to fire off a Runnable a small amount of time after the first note plays, according to the constant NOTE_DELAY. I was a little worried about variances in the timing with this solution at first, but it works exactly as we want.

Take note that we need to keep a final float third variable pointing to the value in currentSecondNote, because the call to incrementScaleDegree will change the instance variable to the next major third in the sequence before the run method is called.

To do this in iOS, we use a dispatch_after with a block of code to run after a specified amount of time.

ios/Tasks/Tasks/PdInterface.m

```objc
- (void) playTaskCompletionCue {
    float secondNote = currentSecondNote;
    [PdBase sendFloat:currentRootNote toReceiver:kMidiNote1];
    dispatch_time_t popTime = dispatch_time(DISPATCH_TIME_NOW,
                                  (int64_t)(kNoteDelay * NSEC_PER_SEC));
    dispatch_after(popTime, dispatch_get_main_queue(), ^(void){
        [PdBase sendFloat:secondNote toReceiver:kMidiNote2];
    });
    [self incrementScaleDegree];
}
```

Note that we need to keep a float secondNote variable pointing to the current value of the major third, because the call to incrementScaleDegree will change the instance variable to the next major third in the sequence before the block is called.

Wiring Up the "Clear Completed" Cue

When the user clears away all of the completed tasks we want to play a sequence of two major thirds followed by two major thirds an octave up. The methods that get called upon clearing all completed tasks first delete all the completed tasks and then call playClearTasksCue. First, the Android version:

android/TasksProject/Tasks/src/main/java/com/programmingsound/tasks/TasksActivity.java
```java
private void clearComplete() {
  Uri tasksUri = TasksContentProvider.CONTENT_URI;
  getContentResolver().delete(
      tasksUri,
      TasksDBHelper.TASK_COMPLETE_COLUMN + "=1",
      null);
  PdInterface.getInstance().playClearTasksCue();
}
```

And now the iOS version:

ios/Tasks/Tasks/TaskListViewController.m
```objc
[self.collectionView performBatchUpdates:^{
    NSIndexSet *deletedSections = [NSIndexSet indexSetWithIndex:sectionToClear];
    [self.collectionView deleteSections:deletedSections];
    [self.pdInterface playTasksClearedCue];
} completion:^(BOOL finished) {
    removingCompleted = NO;
}];
```

These methods use the same techniques as before, but play two notes. They then reset the scale references. Here's the Android code:

android/TasksProject/Tasks/src/main/java/com/programmingsound/tasks/audio/PdInterface.java
```java
public void playClearTasksCue() {
  final int root = ROOT_MIDI_NOTE;
  final int third = ROOT_MIDI_NOTE + scaleDegrees[2];
  PdBase.sendFloat(MIDINOTE1, root);
  PdBase.sendFloat(MIDINOTE2, third);
  handler.postDelayed(new Runnable() {
    @Override public void run() {
      PdBase.sendFloat(MIDINOTE1, root + 12);
      PdBase.sendFloat(MIDINOTE2, third + 12);
    }
  }, NOTE_DELAY);
  resetScaleDegree();
}
```

And here's the iOS code:

ios/Tasks/Tasks/PdInterface.m
```objc
- (void) playTasksClearedCue {
    float root = kRootMidiNote;
    float third = kRootMidiNote + [scaleDegrees[2] integerValue];
    [PdBase sendFloat:root toReceiver:kMidiNote1];
    [PdBase sendFloat:third toReceiver:kMidiNote2];
    dispatch_time_t popTime = dispatch_time(DISPATCH_TIME_NOW,
                                      (int64_t)(kNoteDelay * NSEC_PER_SEC));
```

```
dispatch_after(popTime, dispatch_get_main_queue(), ^(void){
    [PdBase sendFloat:root + 12 toReceiver:kMidiNote1];
    [PdBase sendFloat:third + 12 toReceiver:kMidiNote2];
});
[self resetScaleDegree];
}
```

Now we've wired up all of the sounds we specified in the designs. A few more housekeeping steps need to be wired up, and then we'll be ready to ship it!

Cleaning Up Resources

When the various activity life-cycle methods are called on TasksActivity in Android, we need to do a few things with libpd.

android/TasksProject/Tasks/src/main/java/com/programmingsound/tasks/TasksActivity.java
```
@Override protected void onStart() {
  super.onStart();
  PdAudio.startAudio(this);
}

@Override protected void onStop() {
  PdAudio.stopAudio();
  super.onStop();
}

@Override protected void onDestroy() {
  PdInterface.getInstance().destroy();
  super.onDestroy();
}
```

In onStart and onStop we turn on and off the sound processing, so nothing happens when the activity is not running. In onDestroy, we make sure we let libpd clean a few things up, too.

android/TasksProject/Tasks/src/main/java/com/programmingsound/tasks/audio/PdInterface.java
```
private void cleanup() {
  PdAudio.release();
  PdBase.release();
}
```

On iOS, when the application goes into the background we should shut off the sound processing. We do that in the AppDelegate.

ios/Tasks/Tasks/AppDelegate.m
```
- (void) applicationDidBecomeActive:(UIApplication *)application {
    self.pdAudioController.active = YES;
}
- (void) applicationDidEnterBackground:(UIApplication *)application {
    self.pdAudioController.active = NO;
}
```

Inside applicationDidBecomeActive and applicationDidEnterBackground we toggle the active state of the PdController so that sound processing doesn't take place when the app is in the background.

Wrap-Up

Now we have a native task-management app on two platforms with a little more flavor, using some good sound-design principles, and an embedded sound engine using Pure Data and libpd!

Things to Think About

Here are a few things to consider about the implementation of our app.

What if the dynamic major-scale stepping were done in Pd? Then we wouldn't have to duplicate that logic on both platforms. How would you implement that?

Do users always want sound? Users usually want a way to opt out of the sound apps. Implement some settings to turn sounds off for either app or both apps.

Next Up

In this chapter, we've accomplished a lot. First we went through the specs for a task-management app built on two platforms. Then we talked about what a good sound design might be for those apps. We built a patch that employed FM synthesis to make a configurable patch with an API designed to be used by an app employing libpd.

Then we went through the code for each app and saw how to initialize, configure, and control Pd using libpd's native API for each platform. I hope this has shown you how powerful the combination of Pure Data and libpd can be for your apps and games, and I hope you can make use of them soon!

CHAPTER 10

Your Journey Begins

This book is only the beginning of your journey as a sound designer. I hope you've learned a lot. Whether you've come from a musical background and wondered how you can work with sound programmatically, you've come from a sound-design background and wondered how you can get more control over the sound you work with, or you just have an interest in what sound can do for a digital experience, I hope this book has excited you about the possibilities of procedural audio with Pure Data.

Let's take a quick look at what we've covered in this book and then go over a few possible next steps on your journey as a sound designer.

A Recap

We've covered a lot of ground, from basic skills to advanced techniques, from sound making to sound modeling to integration with production projects. Here are some highlights.

Making Sound with Pure Data

From the basic oscillator to more advanced tools like arrays, tabread4~, and tabosc4~ for samples and wavetables, you've learned a number of ways to make sound with Pure Data. You've also learned to shape sound using filters from raw components like noise, and to construct waves out of simpler waves.

Building Controls

You've seen how to build abstractions, using subwindows with pd and with subpatches in separate files. You've also learned to employ the built-in user-interface controls that come with Pd, such as sliders and toggles. Building on those, we created an envelope and a low-frequency oscillator—time-based controls using delay and line~ objects.

Creating Effects

We've explored how to put everything you've learned into action by creating some realistic effects: wind, waves, a wineglass, a realistic sword fight, and a lightsaber, using a few different techniques. These exercises and examples have given you a taste of what goes into sound design and analyzing and reproducing real-world sounds.

Synthesis Methods

We've delved into a number of synthesis methods, each with particular strengths and characteristics:

- Additive, which adds sine waves together

- Subtractive, which uses filters to shape a spectrum

- Frequency modulation, which produces complex timbres with only a few oscillators

- Wavetable, which uses a precomputed waveform

- Sampling, which builds on a full prerecorded sound

We've used each of these methods in a situation that was well suited to the characteristics of the sound we wanted to create.

UX of Sound Design

We've also taken a step back and looked at the goals of a sound designer and discussed good user-experience (UX) practices for sound. We've seen how sound should be a layer on the UX we're building, and that only very rarely is sound a necessary ingredient (at least in apps, as opposed to games). Done right, sound should enhance the interface subtly and, above all, have a UX goal driving its use.

Using Pure Data for Real-World Projects

Finally, we built some real-world projects for both mobile and web platforms using Pure Data as the sound-creation tool, and the embedded synth engine for the native platforms. We covered libpd and how it works to wrap Pd with a native interface.

That's a solid base on which to build your skills as a sound designer! Now let's look at some of the next steps you could take.

Next Steps

You could take the skills you've developed in this book in a few directions. I'm sure you'd like to dive in and build that music app taking shape in your mind, or replace some static samples with dynamic sounds in the game you've been working on with friends. By all means, build something awesome with Pd! Here are a few other suggested steps to take next to build your skills as a sound designer.

Practice!

It never hurts to practice. Designing sounds just to explore how you might solve a particular case is a great way to enhance your skills. Try to model a bird chirping, the sound of wind in the trees, a gunshot, a car engine, or an ambient musical soundscape that reacts in different ways to input.

Develop Your Critical-Listening Skills

If you already have a musical background, try applying your ear to real-world sounds and ambient noises in the same way you do to music. Pick apart what you hear in the everyday sounds of traffic, or a dinner party, or a crowded subway. What "parts" do you hear? How would you break down and reproduce those sounds?

If you don't have a musical background, try the forgoing suggestions, but consider building up some musical skills too. You don't have to learn an instrument, although that would be fun. Listening to music critically—picking out the different sounds and voices, understanding how they work together—is a great way to train your ears. Developing an understanding of music theory informs how toneful sounds work together, and will help you craft sounds with better effect.

Further Reading

Studying up on sound-design principles is a great next step. Here are some resources for Pd and sound design in general:

- *The Theory and Technique of Electronic Music*,[1] a book by Miller Puckette, creator of Pd

- Programming Electronic Music in Pd[2]

---

1. http://msp.ucsd.edu/techniques.htm
2. http://www.pd-tutorial.com

- Pd mailing lists[3]
- Pd forums at Create Digital Noise[4]
- Pd tutorial videos on YouTube by cheetomoskeeto[5]
- Sound-design topics on Designing Sound[6]

To learn more about libpd, you should check out Peter Brinkmann's book, *Making Musical Apps [Bri12]*.

The effects we've created were varied, interesting, and realistic, but there are many more complex sounds to learn how to create. The best resource for diving deep into sound-design principles as well as applying these to Pure Data is the excellent *Designing Sound [Far12]*, by Andy Farnell,[7] a resource that I have yet to exhaust for my own sound-design development.

Wrapping Up

Sound is a beautiful part of any digital experience, web or native. Designing sound is a worthy pursuit and a continuous source of delight for an audience. I'm glad to be able to introduce this discipline to you, or to further enhance your capabilities as a sound designer. I look forward to hearing what you create with Pure Data!

---

3. http://lists.puredata.info/listinfo
4. http://createdigitalnoise.com
5. http://www.youtube.com/user/cheetomoskeeto
6. http://designingsound.org
7. http://obiwannabe.co.uk

Glossary

Amplitude
> The amount of energy in an audio signal.

Bit Depth
> The number of bits available per sample to store signal data.

Chorusing
> An effect added to a sound by combining it with a delayed copy with a longer duration than flanging. It adds an interesting "doubling" effect and, if modulated, can give the impression of motion.

Flanging
> An effect added to a sound by combining it with a slightly delayed copy. It adds an interesting "whooshing" effect caused by interference in different frequencies in the sound.

Frequency
> The speed of an audio signal, which determines the pitch of a sound.

Frequency-modulation synthesis
> A synthesis method where a modulator operator's frequency changes the frequency of a carrier operator.

Index
> In the context of FM synthesis, the amplitude of the modulator operator; the strength with which the modulator affects the carrier's frequency.

Modulation
> Varying a signal over time.

Noise
> A random signal across a certain frequency range.

Octave

    In music, a note double the frequency of another note, or the distance between two such notes. So called because in Western music scales are generally seven tones, with the sequence repeating on the eighth tone. An Octave is also 12 half steps above a given note.

Operator

    In the context of FM synthesis, an oscillator.

Partial

    One of a set of frequencies that make up a sound.

Period

    The amount of time a wave takes to make one full cycle.

Sample-based synthesis

    A synthesis method that uses longer prerecorded samples of a sound.

Sample rate

    The number of samples per second a digital recording captures.

Scale

    In music, a sequence of steps from a given note to an octave above that note.

Signal

    A digital stream of numbers that encode a sound.

Spectrum

    The distribution of the strength of the frequencies in a signal.

Timbre

    The tonal quality of a sound. Related to spectral content.

Wavetable synthesis

    A synthesis method that uses single-cycle samples of periodic waves.

Bibliography

[Bri12]   Peter Brinkmann. *Making Musical Apps*. O'Reilly & Associates, Inc., Sebastopol, CA, 2012.

[Far12]   Andy Farnell. *Designing Sound*. MIT Press, Cambridge, MA, 2012.

Index

Open GL and Processing

Dive into OpenGL on Android, and explore Processing for game and music development.

OpenGL ES 2 for Android

Android is booming like never before, with millions of devices shipping every day. It's never been a better time to learn how to create your own 3D games and live wallpaper for Android. You'll find out all about shaders and the OpenGL pipeline, and discover the power of OpenGL ES 2.0, which is much more feature-rich than its predecessor. If you can program in Java and you have a creative vision that you'd like to share with the world, then this is the book for you.

Kevin Brothaler
(346 pages) ISBN: 9781937785345. $38
http://pragprog.com/book/kbogla

Rapid Android Development

Create mobile apps for Android phones and tablets faster and more easily than you ever imagined. Use "Processing," the free, award-winning, graphics-savvy language and development environment, to work with the touchscreens, hardware sensors, cameras, network transceivers, and other devices and software in the latest Android phones and tablets.

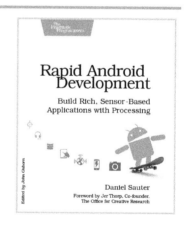

Daniel Sauter
(392 pages) ISBN: 9781937785062. $33
http://pragprog.com/book/dsproc

The Modern Web

Get up to speed on the latest HTML, CSS, and JavaScript techniques.

HTML5 and CSS3 (2nd edition)

HTML5 and CSS3 are more than just buzzwords—
they're the foundation for today's web applications.
This book gets you up to speed on the HTML5 elements
and CSS3 features you can use right now in your current projects, with backwards compatible solutions
that ensure that you don't leave users of older browsers
behind. This new edition covers even more new features, including CSS animations, IndexedDB, and
client-side validations.

Brian P. Hogan
(300 pages) ISBN: 9781937785598. $38
http://pragprog.com/book/bhh52e

Async JavaScript

With the advent of HTML5, front-end MVC, and
Node.js, JavaScript is ubiquitous—and still messy.
This book will give you a solid foundation for managing
async tasks without losing your sanity in a tangle of
callbacks. It's a fast-paced guide to the most essential
techniques for dealing with async behavior, including
PubSub, evented models, and Promises. With these
tricks up your sleeve, you'll be better prepared to
manage the complexity of large web apps and deliver
responsive code.

Trevor Burnham
(104 pages) ISBN: 9781937785277. $17
http://pragprog.com/book/tbajs

The Joy of Math and Healthy Programming

Rediscover the joy and fascinating weirdness of pure mathematics, and learn how to take a healthier approach to programming.

Good Math

Mathematics is beautiful—and it can be fun and exciting as well as practical. *Good Math* is your guide to some of the most intriguing topics from two thousand years of mathematics: from Egyptian fractions to Turing machines; from the real meaning of numbers to proof trees, group symmetry, and mechanical computation. If you've ever wondered what lay beyond the proofs you struggled to complete in high school geometry, or what limits the capabilities of the computer on your desk, this is the book for you.

Mark C. Chu-Carroll
(282 pages) ISBN: 9781937785338. $34
http://pragprog.com/book/mcmath

The Healthy Programmer

To keep doing what you love, you need to maintain your own systems, not just the ones you write code for. Regular exercise and proper nutrition help you learn, remember, concentrate, and be creative—skills critical to doing your job well. Learn how to change your work habits, master exercises that make working at a computer more comfortable, and develop a plan to keep fit, healthy, and sharp for years to come.

This book is intended only as an informative guide for those wishing to know more about health issues. In no way is this book intended to replace, countermand, or conflict with the advice given to you by your own healthcare provider including Physician, Nurse Practitioner, Physician Assistant, Registered Dietician, and other licensed professionals.

Joe Kutner
(254 pages) ISBN: 9781937785314. $36
http://pragprog.com/book/jkthp

Android on Android, 3D for Kids

Script your Android device right on the device, and get your kids writing 3D games in JavaScript.

Developing Android on Android

Take advantage of the open, tinker-friendly Android platform and make your device work the way you want it to. Quickly create Android tasks, scripts, and programs entirely on your Android device—no PC required. Learn how to build your own innovative Android programs and workflows with tools you can run on Android itself, and tailor the Android default user interface to match your mobile lifestyle needs. Apply your favorite scripting language to rapidly develop programs that speak the time and battery level, alert you to important events or locations, read your new email to you, and much more.

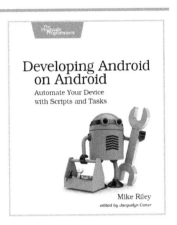

Mike Riley
(232 pages) ISBN: 9781937785543. $36
http://pragprog.com/book/mrand

3D Game Programming for Kids

You know what's even better than playing games? Creating your own. Even if you're an absolute beginner, this book will teach you how to make your own online games with interactive examples. You'll learn programming using nothing more than a browser, and see cool, 3D results as you type. You'll learn real-world programming skills in a real programming language: JavaScript, the language of the web. You'll be amazed at what you can do as you build interactive worlds and fun games. Appropriate for ages 10-99!

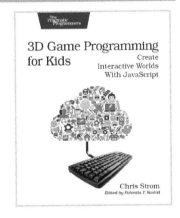

Chris Strom
(250 pages) ISBN: 9781937785444. $36
http://pragprog.com/book/csjava

The Pragmatic Bookshelf

The Pragmatic Bookshelf features books written by developers for developers. The titles continue the well-known Pragmatic Programmer style and continue to garner awards and rave reviews. As development gets more and more difficult, the Pragmatic Programmers will be there with more titles and products to help you stay on top of your game.

Visit Us Online

This Book's Home Page
http://pragprog.com/book/thsound
Source code from this book, errata, and other resources. Come give us feedback, too!

Register for Updates
http://pragprog.com/updates
Be notified when updates and new books become available.

Join the Community
http://pragprog.com/community
Read our weblogs, join our online discussions, participate in our mailing list, interact with our wiki, and benefit from the experience of other Pragmatic Programmers.

New and Noteworthy
http://pragprog.com/news
Check out the latest pragmatic developments, new titles and other offerings.

Save on the eBook

Save on the eBook versions of this title. Owning the paper version of this book entitles you to purchase the electronic versions at a terrific discount.

PDFs are great for carrying around on your laptop—they are hyperlinked, have color, and are fully searchable. Most titles are also available for the iPhone and iPod touch, Amazon Kindle, and other popular e-book readers.

Buy now at *http://pragprog.com/coupon*

Contact Us

Online Orders:	*http://pragprog.com/catalog*
Customer Service:	*support@pragprog.com*
International Rights:	*translations@pragprog.com*
Academic Use:	*academic@pragprog.com*
Write for Us:	*http://pragprog.com/write-for-us*
Or Call:	+1 800-699-7764